Famous Comets

John Hebert

7/21/19-1/13/20

For— Matthew, Barbara's son, who is a great lover of the beauty and majesty of comets.

In memory of—Meryem, Ray, Bunny, Spunky, Guy, my fish, and my birds.

Also for— Cathy, Barbara, Peter, Mabel, Rosie.

Contents of comets—

About comets
1. Brooks
2. Halley
3. Ikeya Seki
4. Encke
5. Honda Mrkos
6. West
7. P/Linear

8. Holmes
9. Arend Roland
10. Schwassman Wachmann
11. 67p/Churyumov Gerasimenko
12. Hyakutake
13. Tempel 1
14. 2060 Chiron
15. 81p/Wild
16. 2011 L4
17. Biela
18. Hale Bopp
19. Borrelly

20. Brorsen
21. Lovejoy
22. Lexells
23. Kohoutek
24. Swift Tuttle
25. Great Comet of 1843
26. Wilson Harrington
27. Wirtanen
28. Pons Winnicke
29. Hartley
30. Gehrels
31. ISON
32. Great Comet of 1910

33. Giacobinai Zinner
34. Stephan Oterma
35. D'Arrest
36. Tempel Tuttle
37. Ikeya Zhang
38. Thatcher
39. 2010 x1 Elenin
40. Skjellerup Maristang
41. Humanson
42. Bennet
43. Mcnaught
44. Morehouse
45. Grigg Skjellerup
46. Tempel 2

47. Schwassmann Wachmann 2

48. Brook 2

49. Finley

50. Faye

51. Wolf 1

52. Tuttle

53. Great Comet Donati

54. Great Comet Tubbutt

55. Great Comet Coggia

56. Great Southern Comet of 1882

57. Great Daylight comet of 1910

58. Flaugergues
59. IRAS Araki Alock
60. Arend Rigaux
61. Pons Gambert
62. Ikeya 1963 1
63. Neujmin
64. Great Comet 1680
65. Pons Brooks
66. Great Comet 1948
67. Blanpain

68. Perrine Mrkos
69. West Kohoutek
70. Honda Mrkos Pajdusakova
71. Great Comet 1769
72. Herschel Rigollet
73. Olbers
74. 1999 S4 Linear
75. Shoemaker Levy
76. Slaughter Burnham
77. The eclipse comet

Pictures of the surface of a comet

Formulas used in orbital calculations

How to Calculate orbits

glossary

About Comets

Comets are small, icy astronomical bodies the come close to the sun and warm up, thus releasing gases which causes tails to form which are always pointed away from the sun. Some comets return to the sun's vicinity repeatedly while others only once. Comets produce comas, which are of great size of extended gases surrounding the nucleus of the comet. The comet produced 2 tails — an ion tail, which always points directly away from the sun, and a dust tail, which is curved. These tails can reach lengths of up to 360 million miles long. Comets can have orbital periods as short as a few years to as long as hundreds of thousands of years. Some comets never return to the solar system.

Comet #1—
Name— Comet Brooks

Picture comet—

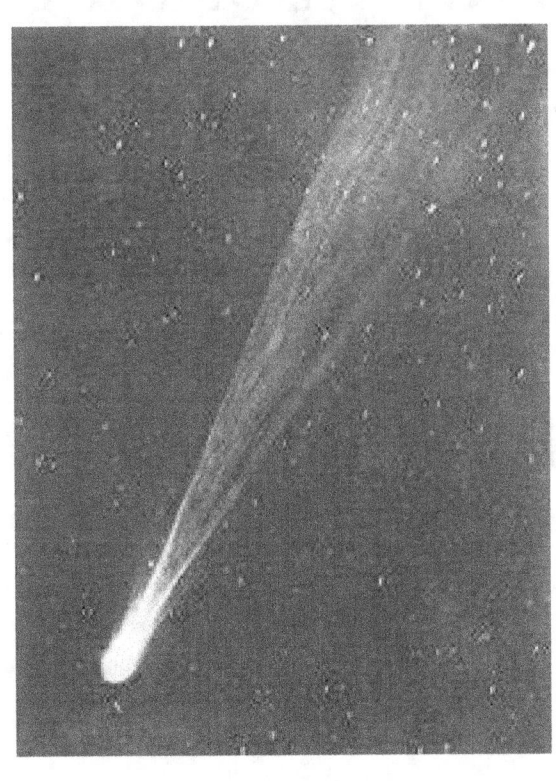

Picture discoverer—n/a

Type of comet—elliptical orbit

Comet Brooks was discovered on July 21, 1911 by William Robert Brooks. It was an apparent magnitude 2 comet with a narrow, straight tail 30 degrees long with a distinct blue color. Another comet, c/1911 53 (Beljawsky) was a 1st magnitude comet in the same part of the sky in mid October 1911.

Orbit formula—

$1 = x^2/153.7^2 + y^2/11.8867^2$

perihelion = .489429 AU

aphelion = 306.9 AU

a = 153.7 AU

b = 11.886715 AU

eccentricity = .997005

1/a = 4.349113307 x 10^-14

period = 3,811 years 6 days

Perihelion statistics—

Distance from sun at its closest= 45 million, 483 thousand 611.92 miles

Velocity at closest distance from sun= 37 miles, 1,195 feet, .852 inches/second

Time of perihelion= 10/28/1911

Aphelion statistics—

Distance from sun at its farthest= 28 billion, 520 million, 828 thousand, 350 miles

Velocity at farthest distance from sun= 318 feet, 1.29 inches/second

Time of aphelion= 2817 AD

Return time= 5722 AD

Statistics of the comet on 7/15/19—

Distance from sun= 14 billion, 700 million, 91 thousand, 310 miles

velocity= 1 mile, 2,374 feet, 7.02 inches/ second

Time since perihelion= 108 years, 259 days (since 7/15/2019)

Returns— 5722 AD

Comet #2—
Name— Comet Halley

Picture comet—

Picture discoverer—

Edmund Halley

Type of comet—elliptical orbit

Comet Halley is a short period comet (20-200 year orbital period). It is the only such comet which can be seen 2 times in a lifetime. It was discovered by Edmund Halley in 1758 and he predicted its return. It was first observed in 240 BC. Comet Halley is a comet originally from the Oort Cloud. It has been in its current orbit for 16,000-200,000 years. Its projected lifespan is 10 million years. Its mass has reduced 80-90% over the last 2-3 thousand orbits. When Halley comes to perihelion, its coma becomes 62,000 miles in diameter and the tail grows to 62 million miles long. It has a diameter of 6 miles, 4,409 feet, a mass of 2.2×10^{14} kg, density of .6 grams/cm^2, escape velocity of 2.8 millimeters/second, rotation rate of 52 hours, 52 seconds, an albedo of .04, and and apparent magnitude of 28.2 as of 2003. Its magnitude in its 1986 return was 2.1 apparent magnitude. On February 12, 1991, comet Halley was 1 billion, 339 million, 200 thousand miles from the sun. On July 29, 2061, its magnitude will be -.3, and in 2134, it will approach earth to 8 million, 37 thousand miles, and appear at magnitude -2. The comet is associated with the Eta Aquarids and Orionid meteor shower in late October. In terms of history, the defeat of Attila the Hum at the ballet of Chalons in 451 was attributed to the comet's appearance. Genghis Khan was inspired to turn his conquests to Europe by the comet's apparition of 1222. Also, the 1301 return of the comet, Giotto de Bondone painted the comet into one of his paintings.

Orbit formula—

$1 = x^2/17.834^2 + y^2/4.5342^2$

perihelion = .586 AU

aphelion = 35.082 AU

a = 17.834 AU

b = 4.5342 AU

eccentricity = .9855

1/a = 3.7218 x 10^-13

period = 75 years 117 days

Perihelion statistics—

Distance from sun at its closest = 54 million, 458 thousand 157.84 miles

Velocity at closest distance from sun= 33 miles, 4,787 feet/second

Time of perihelion= 1/20/1986

Aphelion statistics—

Distance from sun at its farthest= 3 billion, 263 million, 401 thousand, 440 miles

Velocity at farthest distance from sun= 3,565 feet, 10.81 inches/second

Time of aphelion= 2024

Statistics of the comet on 10/24/19—

Distance from sun= 3 billion, 136 million, 245 thousand, 790 miles

velocity= 2 mile, 2,133 feet, 5.18 inches/ second

Time since perihelion= 35 years, 215 days

Returns— 2061 AD

Comet #3—

Name— Comet Ikeya Seki

Picture comet—

Picture discoverer—

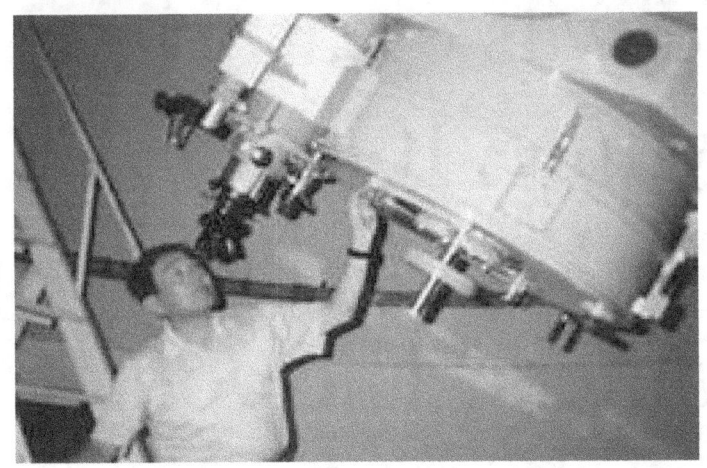

Ikeya   Seki

Type of comet—elliptical orbit

Discovered independently by Kaoro Ikeya and Tsutomu Seki in September 18, 1965, Comet Ikeya Seki was one of the brightest comets in 1,000 years (apparent magnitude -10), which was at its brightest on 10/14/65, and was clearly visible at daytime. It is a member of the Kreutz family of sun grazer comets. The comet made its closest approach to earth at a distance of 83 million, 700 thousand miles. The comet of 1106 broke up into 2 fragments, and the larger fragment is the comet Ikeya Seki. The comet broke up again into 3 fragments before 1965 perihelion. In early 1966, it faded from view as it orbited away from the sun.

Orbit formula—
$1 = x^2/91.6^2 + y^2/1.1943^2$

perihelion= .007786 AU (fragment A) (10/21/65)

aphelion= 183.192214 AU

a= 91.6 AU
b= 1.1943 AU

eccentricity= .999915
1/a= 7.297534229 x 10^-14

period= 876 years 256 days

Perihelion statistics—

Distance from sun at its closest= 723 thousand 568.5 miles

Velocity at closest distance from sun= 296 miles, 3,051 feet, 4.2 inches/second

Time of perihelion= 10/21/65

Aphelion statistics—

Distance from sun at its farthest= 17 billion, 24 million, 417 thousand, 370 miles

Velocity at farthest distance from sun= 66 feet, 6.66 inches/second

Time of aphelion= 2403 AD

Statistics of the comet on 11/14/19—

Distance from sun= 5 billion, 386 million, 458 thousand miles

velocity = 3 mile, 654 feet/second

Time since perihelion = 54 years, 24 days

Returns — 2842 AD

Comet #4 —

Name — Comet Encke

Picture comet —

Picture discoverer—

Encke

Type of comet—elliptical orbit

Comet Encke is the shortest period comet which is

reasonably bright. It was discovered by Pierre Mechain on January 17, 1786, and its orbit was calculated by Johann Encke. It has a diameter of 2,976 miles, and its albedo is low where it reflects only 4.6% of the light it receives. The comet has come as close as 16 million 90 thousand miles to earth, and it usually comes close every 33 years. It loses 2 days in its period of orbit each successive period around the sun. Encke is the source of the

Taurid meteor shower (north and south Taurids) in November and both Taurus in late June and early July. It has been postulated that Comet Encke was the comet whose fragment of it was the cause of the Tunguska event in 1908 in Russia where an object entered the earth's atmosphere and flattened trees over miles range of area. Its next perihelion is on June 25, 2020.

Orbit formula—

$1 = x^2/2.2178^2 + y^2/1.1786^2$

perihelion= .3302 AU

aphelion= 4.11 AU

a= 2.2178 AU

b= 1.1786 AU

eccentricity= .8471

1/a= 2.9928 x 10^-12

period= 1,204 days

Perihelion statistics—

Distance from sun at its closest= 30 million, 686 thousand miles

Velocity at closest distance from sun= 43 miles, 4,366 feet, 4.6 inches/second

Time of perihelion= 3/10/2017

Aphelion statistics—

Distance from sun at its farthest= 381 million, 950 thousand, 500 miles

Velocity at farthest distance from sun= 3 miles, 3,431 feet, 1.634 inches/second

Time of aphelion= 11/23/2018

Returns— 2020

Comet #5—
Name— Comet Honda Mrkos

Picture comet—

Picture discoverer—

Mrkos

Type of comet—elliptical orbit

Comet Honda Mrkos was discovered on 12/3/1948. It is .81 miles in diameter. On 1/16/1996, it had an apparent magnitude of 7 and was 4.3 degrees from the sun. On 8/15/2001, its distance from the earth was 5.58 million miles, and by late September was magnitude 7.3. Its last perihelion was on 12/31/2016. By February 4, 2017, it was magnitude 7 and had a coma 62,000 miles in diameter. By February 11, 2017, its distance from the earth was 7 million, 732 thousand miles.

Orbit formula—

$1 = x^2/3.0205^2 + y^2/1.708746^2$

perihelion= .5296 AU (12/31/2016, 4/26/2022)

aphelion= 5.511 AU

a= 3.0205 AU

b= 1.708746 AU

eccentricity= .8246

1/a= 2.210073118 x 10^-12

period= 5.25 years

Perihelion statistics—

Distance from sun at its closest= 49 million, 216 thousand 783 miles

Velocity at closest distance from sun= 34 miles, 1,851 feet, 10.74 inches/second

Time of perihelion= 12/31/2016

Aphelion statistics—

Distance from sun at its farthest= 512 million, 148 thousand, 208 miles

Velocity at farthest distance from sun= 3 miles, 1,734 feet, 9.08 inches/second

Time of aphelion= 11/23/2018

Returns— 4/26/2022

Comet #6—
Name— Comet West

Picture comet—

Picture discoverer— n/a

Type of comet—elliptical orbit

Comet West was a great comet of 1975 with was discovered 8/10/1975 by Richard West. It reached perihelion on 2/25/1975 and had a minimum elongation of 6.4 degrees with magnitude -3. It could be seen in full daylight. Its period of orbit was 254,000-580,000 years, or even 6.5 million years. Comets with millions of years orbits are unstable because of perturbations from passing stars and galactic tides. Before perihelion, it had an orbit of 254,000 years, then at 18 million, 636 thousands miles from the sun, it split into 4 parts — March 7th into 2 parts and March 18th into 2 more fragments. Since 2003, comet West was 4 billion, 650 million miles from the sun.

Orbit formula—
$1 = x^2/35{,}000.0985^2 + y^2/271.11^2$

perihelion= .197 AU

aphelion= 70,000 AU

a= 35,000.0985 AU

b= 271.11 AU

eccentricity= .99997

1/a= 1.909876669 x 10^-16

period= 654,790 years

Perihelion statistics—

Distance from sun at its closest= 18 million, 307 thousand, 602.43 miles

Velocity at closest distance from sun= 58 miles, 5,078 feet, 10.249 inches/second

Time of perihelion= 2/25/1975

Aphelion statistics—
Distance from sun at its farthest= 6 trillion, 505 billion, 239 million, 441 thousand miles

Velocity at farthest distance from sun= 10.5136 inches/second

Time of aphelion= 329,370 AD

Returns— 656,765 AD

Comet #7—

Name— Comet P/Linear

Picture comet—

Picture discoverer— n/a

Type of comet—elliptical orbit

Comet P/Linear was discovered on 2/3/2004. It is 1.49 miles in diameter and is peanut shaped.

This comet is probably evolving into an extinct comet because of its extremely low level of activity for its size. On 5/29/2014, it was 5 million, 152 thousand miles from earth with a magnitude of 12. It was the 9th closest comet to earth on May 24th when it passed 576,000 miles of us. It produces a meteor shower with a peak of 10-15 meteors/hour.

Orbit formula—
$1 = x^2/2.961^2 + y^2/2.1912^2$

perihelion= .9695 AU (4/15/2009, 6/12/2019)

aphelion= 4.592 AU

a= 2.961 AU

b= 2.1912 AU

eccentricity= .67258

1/a= 2.257543787 x 10^-12

period= 5 years, 33 days

Perihelion statistics—

Distance from sun at its closest= 90 million, 97 thousand, 566.26 miles

Velocity at closest distance from sun= 24 miles, 1,614 feet, 5.5 inches/second

Time of perihelion= 6/12/19

Aphelion statistics—

Distance from sun at its farthest= 460 million, 199 thousand, 224.5 miles

Velocity at farthest distance from sun= 4 miles, 4,010 feet, 5.56 inches/second

Time of aphelion= 12/28/2021

Returns— 2024 AD

Comet #8 —
Name — Comet Holmes
Picture comet —

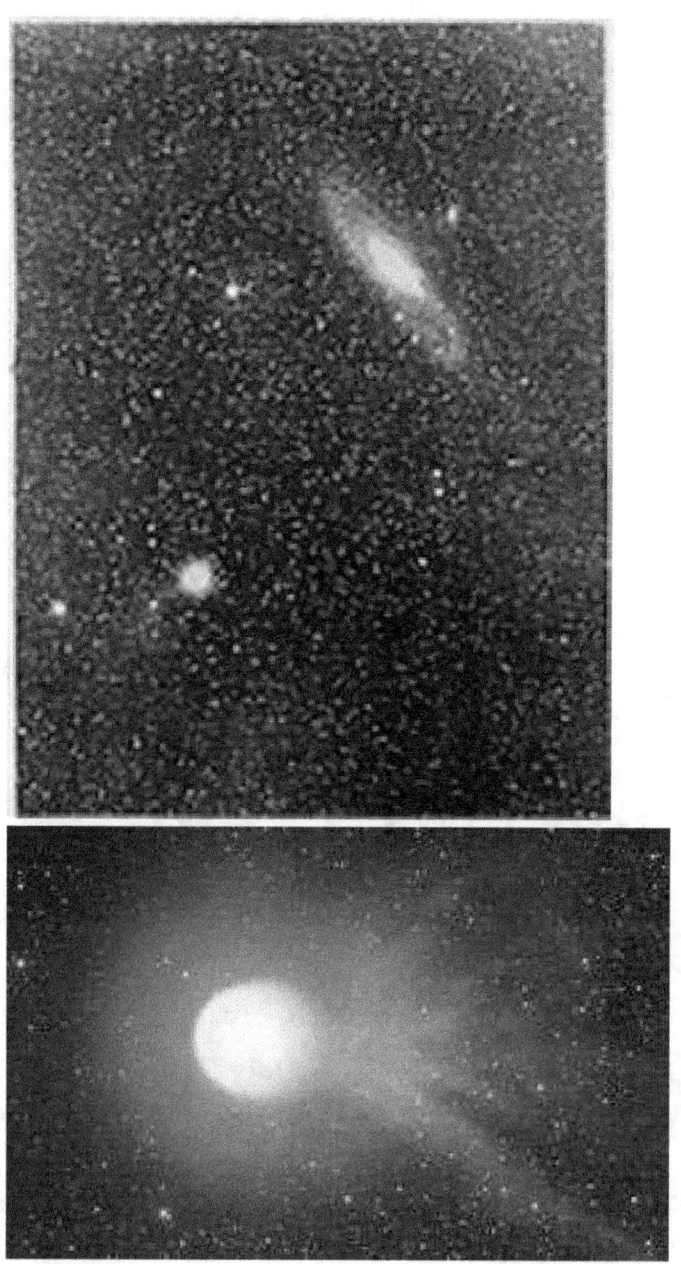

Picture discoverer— n/a

Type of comet—elliptical orbit

Comet Holmes was discovered by Edwin Holmes on 11/6/1892. It has a size of 2 miles, 595 feet in diameter. In 2007, it brightened a half a million fold and was the largest outburst of any known comet. At 184 million miles from the sun, the coma was bigger than the sun (the size of the moon) and it was visible to the naked eye. In 2007 (11/23-24/2007), it went from magnitude 17 to 2.8 in 42 hours. It was seen as bright yellow color in the constellation of Perseus and on October 25th, it was brighter than all the stars there except two. The comet was visible until February 2008 at magnitude 5 in Perseus. By January 2015, there were outbursts again of magnitude 3-4 and needed a large telescope to see the comet.

Orbit formula—
$1 = x^2/3.618414^2 + y^2/3.262373754^2$

perihelion = 2.053218 AU
(5/4/2017, 2/19/2021)

aphelion = 5.18361 AU

a = 3.618414 AU

b = 3.262373754 AU

eccentricity = .432564

1/a = 1.847380414 × 10^-12

period = 1,204 days

Perihelion statistics—

Distance from sun at its closest = 190 million, 809 thousand, 639 miles

Velocity at closest distance from sun = 15 miles, 2,413 feet, 7.7 inches/second

Time of perihelion = 5/4/2017

Aphelion statistics—

Distance from sun at its farthest = 481 million, 723 thousand, 203 miles

Velocity at farthest distance from sun= 6 miles, 647 feet, .885 inches/second

Time of aphelion= 8/12/2020

Returns— 2/19/2022

Comet #9—

Name— Comet Arend Roland

Picture comet—

Picture of discoverer—

Arend

Type of comet—hyperbolic orbit

Comet Arend Roland was discovered by astronomer Sylvian Arend and George Roland on 11/18/1956. In early December 1956, it was 158 million miles from the sun. When it was 158 million, 100 thousand miles from the earth, it reached magnitude 7.5-8, and the tail was 1.5 degrees long. It reached perihelion on 4/8/1957 and produced a prominent display in April in the northern hemisphere of magnitude -1. The tail was 5 degrees long on April 22nd and there was an anti tail 12 degrees long on the 25th of April that split into 3 beams on April 29th, then disappeared. The comet emitted 7.5×10^4 kg/second of gas molecules with a total release of $3 \times 10^6 - 5 \times 10^{10}$ kg. This comet may have originated from the Oort cloud.

Orbit formula— $1 = x^2/(-1,316.8333^2) - y^2/28.85210381^2$

perihelion= .31 AU (4/8/1957)

aphelion= n/a
a= -1,316.8333 AU
b= 28.85210381 AU

eccentricity= 1.00024

period= n/a (ejection from solar system)

Perihelion statistics—

Distance from sun at its closest= 29 million, 370 thousand, 227 miles

Velocity at closest distance from sun= 46 miles, 2,926 feet, 11.58 inches/second

Time of perihelion= 4/8/1957

Velocity at infinity— 2,692 feet, 6.23 inches/second

Note— This comet will not return to the solar system again.

Comet #10—

Name— Comet Schwasson Wachmann

Picture comet—

Picture discoverer— n/a

Type of comet—elliptical orbit

Comet Schwasson Wachmann was discovered on 5/2/1930. During that time, there was a meteor shower of 100 meteors/

minute (Tau Herculid) which occurs every 16 years. The comet is in the process of disintegrating. This breakup began when it re-entered the sun's vicinity in 1995 and broke up into 4 pieces. In March 2006, there were at least 8 fragments. In April of 2006, the Hubble space telescope recorded the disintegration. There are now 66 separate objects left from the comet. The comet was originally 3,608 feet in diameter. On May 12, 2006, the comet passed earth at 7 million 400 thousand miles distance.

Orbit formula—

$1 = x^2/3.063^2 + y^2/2.2163^2$

perihelion= .9426 AU
(3/16/2017, 8/25/2022)

aphelion= 5.184 AU

a= 3.063 AU

b= 2.2163 AU

eccentricity= .6923

1/a= 2.182366031 x 10^-12

period= 1,956 days

Perihelion statistics—

Distance from sun at its closest= 87 million, 597 thousand, 696 miles

Velocity at closest distance from sun= 24 miles, 4,196 feet, 5.67 inches/second

Time of perihelion= 3/16/2017

Aphelion statistics—

Distance from sun at its farthest= 481 million, 759 thousand, 447 miles

Velocity at farthest distance from sun= 4 miles, 3,678 feet, 2.07 inches/second

Time of aphelion= 11/20/2019

Returns— 8/25/2022

Comet #11—

Name— Comet 67p/ Chuyumov- Gerasimenko

Picture comet—

Picture discoverer— n/a

Type of comet—elliptical orbit

Comet 67/Chuyumov-Gerasimenko is a Jupiter family comet from the Kupier belt. It was first observed by the Soviets on 9/20/1969. Its size is 2.67 x 2.54 miles and has a rotation rate of 12 hours and 24 minutes. It has a volume of 4.5 cubic miles, a mass of 9.982×10^{12} kg, mean density of .533 grams per cubic centimeter, escape velocity of 3.28 feet/

second, an albedo of .06, and a temperature of -135 to -45 degrees Fahrenheit. Also, its gravity is .00328 feet/second, it has no magnetic field, has 16 organic compounds on it, and the amino acid glycine was detected, along with oxygen and nitrogen gases. The comet loses 3.3 feet orbit distance per orbit. It has a mass of 10 billion tons and there were a number of collapsing cliffs observed there. The Rosetta spacecraft, which on May 2014 reduced its velocity to 1,700 mph was further reduced to a relative 2 mph and entered orbit on 9/10/2014. Its lander Philae landed on it on 11/12/2014. The lander weighs 220 pounds.

Orbit formula—

$1 = x^2/3.463^2 + y^2/2.65817^2$

perihelion= 1.2432 AU (8/13/2015)

aphelion= 5.6829 AU
a= 3.463 AU
b= 2.65817 AU

eccentricity= .641

1/a= 1.930287945 x 10^-12

period= 2,352 days

Perihelion statistics—

Distance from sun at its closest= 115 million, 533 thousand, 75 miles

Velocity at closest distance from sun= 21 miles, 1,375 feet, 9.24 inches/second

Time of perihelion= 8/13/2015

Aphelion statistics—

Distance from sun at its farthest= 528 million, 123 thousand, 217.4 miles

Velocity at farthest distance from sun= 4 miles, 3,436 feet, 16.14 millimeters/ second

Time of aphelion= 11/3/2021

Returns— 2025 AD

Comet #12—

Name— Comet Hyakutake

Picture comet—

Picture discoverer—

Yuji Hyakutake

Type of comet—elliptical orbit

Comet Hyakutake was discovered on 1/31/1996 by Yuji Hyakutake. It is one of closest of comets in 200 years, passing by earth at a distance of 13 million, 20 thousand miles. It was widely seen around the world. Its greatest magnitude was -1 on 5/1/1996. The comet has the longest tail of any comet known, 360 million miles long (80 degree tail), almost 2 times longer than the great comet of 1843. It was magnitude 0 (visible to the naked eye as bluish green in color) and its coma was 4 times the apparent size of the moon. the velocity of the comet had it cover 1/5 a degree every 30 minutes. Perihelion was on May 1, 1996. On May 25th, it was .1 AU from the earth (9 million 300 thousand miles) and only 4 comets in the 20th century have been closer. The comet had a period of 17,000 years, but perturbations with the planets increased the period to 70,000 years. By May 4th, the tail was

5 degrees long. Its size is 1 mile, 2,922 feet in diameter, in March began dust emissions of 2×10^3 kg/second, at perihelion 3×10^3 kg/second, and there were x-ray emissions coming from the comet's coma.

Orbit formula—
$1 = x^2/1,700^2 + y^2/24.682^2$

perihelion= .2301987 AU (5/1/1996)

aphelion= 3,399.8 AU

a= 1,700 AU

b= 24.682 AU

eccentricity= .9998946

1/a= 3.9044x10^-16

period= 70,000 years

Perihelion statistics—

Distance from sun at its closest= 21 million, 592 thousand, 828 miles

Velocity at closest distance from sun= 54 miles, 2,875 feet, 10.4 inches/second

Time of perihelion= 5/1/1996

Aphelion statistics—

Distance from sun at its farthest= 315 billion, 750 million miles

Velocity at farthest distance from sun= 2,249 feet, .97 inches/second

Time of aphelion= 36,996 AD

Returns— 71,996 AD

Comet #13—
Name— Comet Tempel 1
Picture comet—

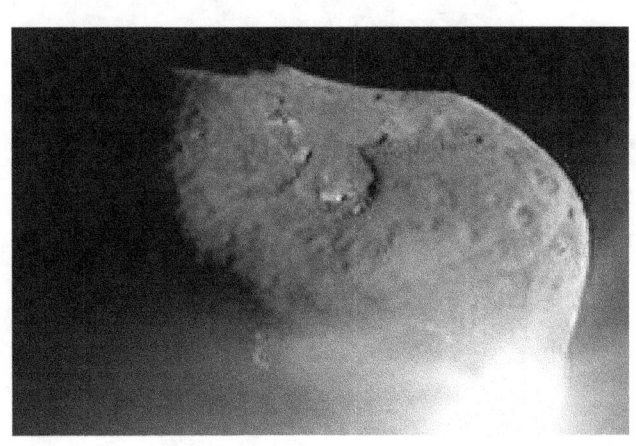

Picture discoverer—

Tempel

Type of comet—elliptical orbit

Comet Tempel 1 is a Jupiter family comet discovered by Wilhelm Temple on 4/3/1867. It has a diameter of 3.6082 miles, rotates in 40 hours and 42 minutes, and has a low albedo of 4%. The comet was the target of the space probe Deep Impact, which impacted with the comet and raised dust 328-820 feet high and made a 490 foot diameter crater. The impact had the energy of 9 billion joules (4.8 tons of TNT) by its 6.346 miles per second impact velocity. The craft weighed 816 pounds. Temple 1 was revisited on 2/14/2011 by the Stardust spacecraft.

Orbit formula—

$1 = x^2/3.145^2 + y^2/2.706^2$

perihelion = 1.542 AU (8/2/2016, 3/4/2022)

aphelion = 4.748 AU

a = 3.145 AU

b = 2.706 AU

eccentricity = .5096

$1/a$ = 2.125464914 × 10^{-12}

period = 2,038 days

Perihelion statistics—

Distance from sun at its closest= 143 million, 301 thousand, 132 miles

Velocity at closest distance from sun= 18 miles, 1,637 feet, 10.225 inches/second

Time of perihelion= 8/2/2016

Aphelion statistics—

Distance from sun at its farthest= 441 million, 241 thousand, 98 miles

Velocity at farthest distance from sun= 5 miles, 4,997 feet, 10.85 inches/second

Time of aphelion= 8/19/19

Returns— 3/4/2022

Comet #14—

Name— 2060 Chiron

Picture comet—

Picture discoverer—

Charles Kowal

Type of comet—elliptical orbit

2060 Chiron was discovered in 1977 by Charles Kowal. It is a minor planet that exhibits cometary behavior, and is located between Saturn and Uranus (a Centaur body orbiting between the asteroid belt and the copier belt.) its angular diameter is less than .035 seconds (66.97 x 72.5 miles), rotates in 5 hours and 55 minutes, has an average orbital velocity of 4.8144 miles/second, and has an absolute magnitude of 5.8+/ 1.27 with apparent magnitude of 18.93 (15.6 at perihelia opposition).

Orbit formula—
$1 = x^2/13.648^2 + y^2/12.6113^2$

perihelion= 8.4311 AU

aphelion= 18.865 AU
a= 13.648 AU
b= 12.6113 AU

eccentricity= .3823

1/a= 4.89785108 x 10^-13

period= 18,417 days

Perihelion statistics—

Distance from sun at its closest= 783 million, 518 thousand, 918 miles

Velocity at closest distance from sun= 7 miles, 2,601 feet, 8.215 inches/second

Time of perihelion= 1971

Aphelion statistics—

Distance from sun at its farthest= 1 billion, 753 million, 102 thousand, 29 miles

Velocity at farthest distance from sun= 4 miles, 3,436 feet, 16.14 inches/second

Time of aphelion= 2014/2015

Returns— 2027 AD

Comet #15—

Name— Comet 81p/Wild

Picture comet—

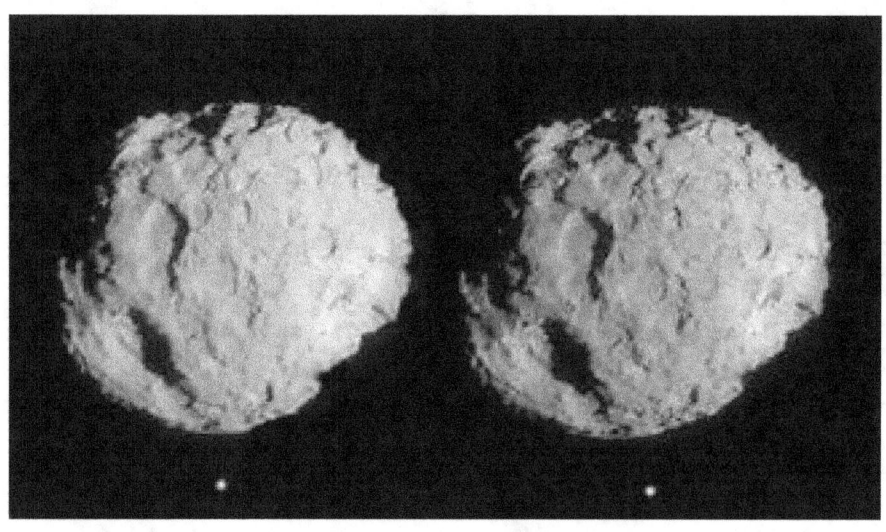

Picture discoverer—

Paul Wild

Type of comet—elliptical orbit

Comet 81P/wild was discovered by Paul Wild on 1/6/1978. It has a diameter of 3.1068 miles, a density of 37 lbs ft^3, and has a mass of 2.3 x 10^13 kgs. For most of 4.5 billion years, the comet was more distant and had a circular orbit, but in September 1974, it passed within 621,000 miles of Jupiter and brought it into the inner solar system and changed its orbital period from 43 to 6 years. On 2/7/1999, Stardust space probe was launched and flew by the comet on 1/2/2004 and collected samples from its coma and returned them to earth along with interstellar dust. The probe returned to earth in Utah on 1/15/2006. The returned dust contains a wide range of organic compounds.

Orbit formula—

$1 = x^2/3.35^2 + y^2/2.90728^2$

perihelion = 1.592 AU

aphelion= 5.308 AU

a= 3.35 AU

b= 2.90728 AU

eccentricity= .5384

1/a= 1.937561494 x 10^-12

period= 6 years, 149 days

Perihelion statistics—

Distance from sun at its closest= 147 million, 947 thousand, 731.3 miles

Velocity at closest distance from sun= 18 miles, 1,012 feet, 5.61 inches/second

Time of perihelion= 7/30/2016, 12/15/2022

Aphelion statistics—

Distance from sun at its farthest= 493 million, 283 thousand, 13.6 miles

Velocity at farthest distance from sun= 5 miles, 2,408 feet, 6 inches/second

Time of aphelion= 10/13/2019

Returns— 12/15/2022

Comet #16—

Name— Comet 2011 L4

Picture comet—

Picture discoverer— n/a

Type of comet—hyperbolic orbit

Comet 2011 L4 is a non periodic comet discovered in June of 2011. On June 11th, it was magnitude 19 at 734 million, 700 thousand miles from the sun, and may 12th, it was magnitude 13.5. October 12th, the coma grew to 75,000 miles in diameter. February 7, 2013, it was magnitude 6. It was visible at perihelion March 10, 2013 at magnitude 1. The comet was closest to earth on March 13th at 101 million, 37 thousand miles distance. March 17-18, it was near the 2.8 magnitude star Al Genib (Gamma Pegasi), April 22 near Beta Cassiopeiae, and May 12-14 near Gamma

Cephei. It probably took millions of years for this comet to come from the Oort Cloud. The comet is a very young baby class comet with a photometric age of less than 4 comet years.

Orbit formula— $1 = x^2/(-3{,}466.78^2) - y^2/45.731^2$

perihelion= .30161 AU (3/10/2013)

aphelion= n/a
a= -3,466.78 AU
b= 45.731 AU

eccentricity= 1.000087

period= n/a

Perihelion statistics—

Distance from sun at its closest= 28 million, 29 thousand, 218 miles

Velocity at closest distance from sun= 33 miles, 3,678 feet, 1.4 inches/second

Time of perihelion= 3/10/2013

Velocity at infinity= 1,659 feet, 5.25 inches/second

Note-This comet will not return again.

Comet #17—

Name— Comet Biela

Picture comet—

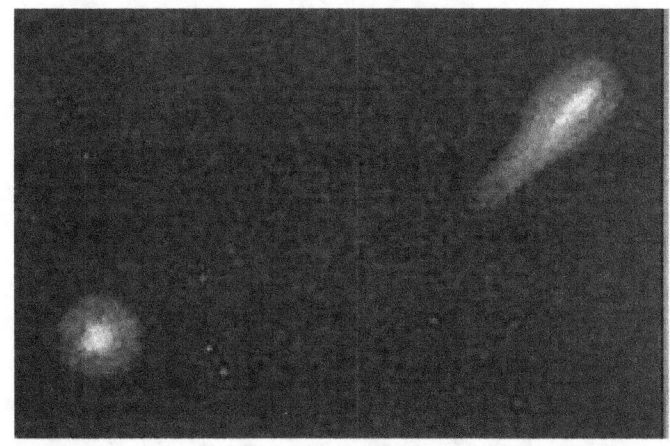

Picture discoverer— n/a
Type of comet—elliptical orbit

Orbit formula—
$1 = x^2/3.5253^2 + y^2/2.30797^2$

perihelion= .8606 AU (9/24/1852)

aphelion= 3.19 AU
a= 3.5253 AU
b= 2.30797 AU

eccentricity= .7559

1/a= 1.89175405 x 10^-12

period= 2,416 days

Perihelion statistics—

Distance from sun at its closest= 79 million, 977 thousand, 272.33 miles

Velocity at closest distance from sun= 26 miles, 2,283 feet, 1.798 inches/second

Time of perihelion= 9/24/1852

Aphelion statistics— n/a (disintegrated comet)

Comet #18—
Name— Comet Hale Bopp

Picture comet

Picture discoverer—

Bopp   Hale

Type of comet—elliptical orbit

Comet Hale Bopp was perhaps the most observed comet in history and the brightest in decades. 60% of Americans saw it by 4/9/1997. It was discovered by Alan Hale and Thomas Bopp on 7/27/1995. It could even be seen in the cities in January 1997. It had magnitude 0 for 8 weeks and was visible for 18 months. The tail was 2 times longer than the great comet of 1811 and the 3rd tail was a sodium tail 31 million miles long between the other 2 tails. At 1 billion, 209 million miles from the sun, it was 100 times brighter than Halley's comet and 6 times its size. On the discovery date, it was 976 million, 500 thousand miles from the sun near the globular cluster M1470 in Sagittarius. At 669 million, 600 thousand miles from the sun, it was between Jupiter and Saturn and by far the greatest distance a comet has been discovered by amateurs, and it already had a coma. In December 1996-January 1997, it was close enough to be observable. At magnitude 2 in February 2017, a pair of tails, one a blue gas tail pointed away from the sun and a yellow dust tail curved along its orbit appeared. The comet came closest to earth on 3/22/1997 at 122 million, 295 thousand miles. On April 1, 1997, the perihelion comet was brighter than any night sky object (magnitude -1.5) except Sirius and its tail was 4-45 degrees long. It was visible all night in the northern hemisphere, and was visible for 569 days until December 1997. Ten years after perihelion, it was 2 billion, 390 million, 100 thousand miles from the sun, and 8/7/2012 it was 2 billion

885 million, 100 thousand miles from the sun. By 2020, it can be seen by telescope at magnitude 30. A previous perihelion 4200 years ago in July 2215BC brought it 130 million, 200 thousand mile to the earth. A near collision in early June 2215 BC, a near collision with Jupiter caused a dramatic change in its orbit so that it was almost perpendicular to the plane of the ecliptic and the orbit shortened from 525 to 370 AU at aphelion. There is a 2.5 x

10^{-9} chance the comet will hit the earth per orbit and would cause a 4.4 megaton impact event. The comet is 24.8-49.6 miles in diameter, has an albedo of .01-.07, a density of .6 grams/cm^3, a mass of 1.3×10^{16} kg, and a velocity of 32.55 miles/second.

Orbit formula—
$1 = x^2/186^2 + y^2/18.4157^2$

perihelion= .914 AU (4/1/1997)

aphelion= 370.8 AU

a= 186 AU

b= 18.4157 AU

eccentricity= .995086021

1/a= 3.568527868 x 10^-14

period= 2,536 years, 356 days

Perihelion statistics—

Distance from sun at its closest= 84 million, 939 thousand, 857 miles

Velocity at closest distance from sun= 27 miles, 1,796 feet, 3.17 inches/second

Time of perihelion= 4/1/1997

Aphelion statistics—

Distance from sun at its farthest= 34 billion, 459 million, 183 thousand, 300 miles

Velocity at farthest distance from sun= 726 feet, 7.15 inches/second

Time of aphelion= 3,266 AD

Returns— 4,534 AD

Comet #19—

Name— Comet Borrelly

Picture comet—

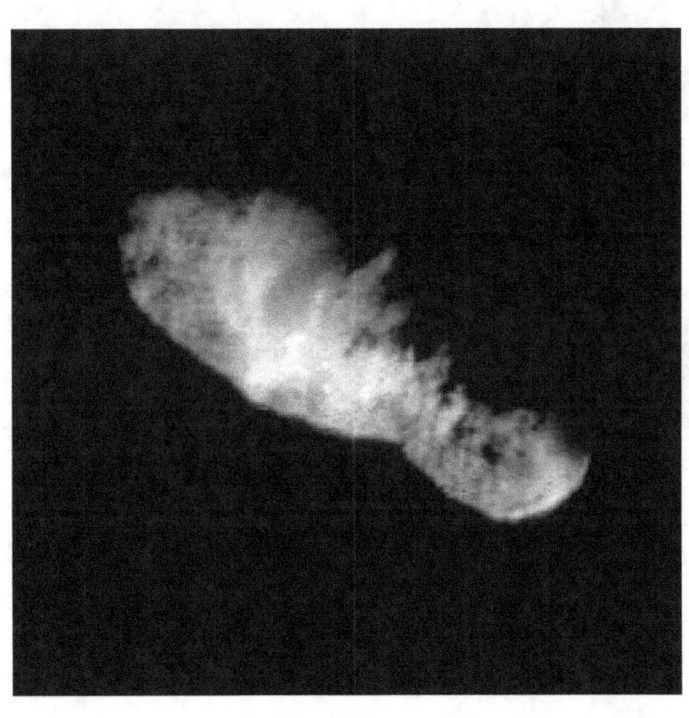

Picture discoverer—

Alphonse Borrelly

Type of comet—elliptical orbit

Comet Borrelly was discovered by Alphonse Borrelly on 12/28/1904. It is 1 mile, 4,503 feet in diameter, shaped like a bowling pin, has a mass of 2 x 10^13 kg, a density of .03 grams/cm^3, and an albedo of .03. the comet was visited by the spacecraft Deep Space 1 on 9/22/2001.

Orbit formula—

1=x^2/3.59^2+y^2/2.8053^2

perihelion= 1.35 AU

aphelion= 5.83 AU

a= 3.59 AU

b= 2.8053 AU

eccentricity= .624

1/a= 1.862001933 x 10^-12

period= 6 years, 292 days

Perihelion statistics—

Distance from sun at its closest= 125 million, 458 thousand, 189 miles

Velocity at closest distance from sun= 20 miles, 1,562 feet, 10.377 inches/second

Time of perihelion= 5/28/2015

Aphelion statistics—

Distance from sun at its farthest= 541 million, 793 thousand, 618 miles

Velocity at farthest distance from sun= 4 miles, 3,694 feet, 8.81 inches/second

Time of aphelion= 10/22/2018

Returns— 2022 AD

Comet #20—
Name— Comet Brörsen

Picture comet—

Picture discoverer—

Theodore Brorsen

Type of comet—elliptical orbit

Comet Brorsen was discovered on 2/26/1846 by Danish Theodore Brorsen. It is a Jupiter family comet. Perihelion occurred February 25th. The comet's coma was 3-4 arc minutes on march 9th and 8-10 arc minutes march 22nd. The comet past earth march 27th at a distance of 48 million, 36 thousand miles. Its orbital period was well known by June

1857. In 1879, it was observed for 4 months, the longest to date. It is currently considered a lost comet.

Orbit formula—

$1 = x^2/3.1^2 + y^2/1.8188^2$

perihelion= .5898 AU

aphelion= 5.61 AU

a= 3.1 AU

b= 2.8188 AU

eccentricity= .8098

1/a= 2.156318437 x 10^{-12}

period= 1,993 days

Perihelion statistics—

Distance from sun at its closest= 54 million, 811 thousand, 289 miles

Velocity at closest distance from sun= 32 miles, 2,191 feet, 3.07 inches/second

Time of perihelion= 3/31/1879

Note— currently considered a lost comet.

Comet #21—

Name— Comet Lovejoy

Picture comet—

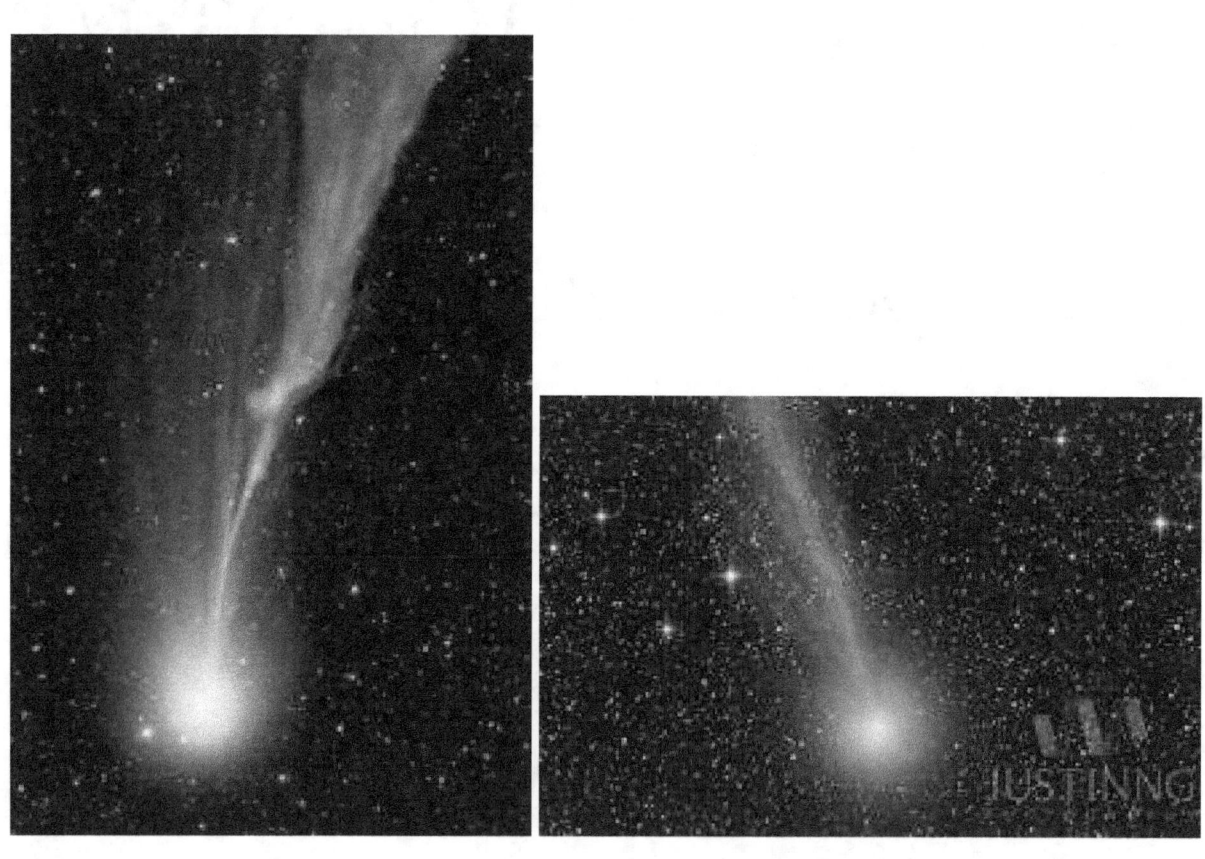

Picture discoverer—

Lovejoy

Type of comet—elliptical orbit

Comet Lovejoy is a long period Kreutz sun grazing comet discovered by Terry Lovejoy in 2011 in Australia. The comet moved rapidly at 13th magnitude when it was discovered. It passed through the sun's corona on 12/16/2011 intact but greatly affected by the passage. Its apparent magnitude was -3 to -4 but was nearly invisible due to the proximity to the sun when the comet was at peak brightness. The nucleus was 1,640 feet across, but is now 328-656 feet across due to mass being burned off during its very close approach to the sun.

Orbit formula—
$1 = x^2/78.68^2 + y^2/.930938^2$

perihelion= .00555 AU (12/16/2011)

aphelion= 157.36 AU

a= 78.68 AU

b= .930938 AU

eccentricity= .99993

1/a= $8.43602165 \times 10^{-14}$

period= 697 years, 331 days

Perihelion statistics—

Distance from sun at its closest= 515 thousand, 772 miles

Velocity at closest distance from sun= 351 miles, 1,623 feet, 7.19 inches/second

Time of perihelion= 12/16/2011

Aphelion statistics—

Distance from sun at its farthest= 14 billion, 633 million, 289 thousand, 490 miles

Velocity at farthest distance from sun= 924 feet, 10.47 inches/second

Time of aphelion= 2360 AD

Returns— 2710 AD

Comet #22—
Name— Comet Lexells

Picture comet—

Picture discoverer—

Charles Messier
Type of comet—elliptical orbit

Comet Levels was discovered 6/14/1770 by Charles Messier. It passed closer to earth than any other comet in history at 1 million, 395 thousand miles. However, comet p/1999 j6 passed by earth at 1 million, 116 thousand miles, though unobserved. Comet Lexells is considered lost. When it was discovered, it was passing in the constellation Sagittarius. On June 24th, its coma was 27 arc minutes large and it was magnitude 2. It crossed 42 degrees of the sky in 24 hours. The last observation of Levels was on 8/3/1770. It came from outside of the solar system on a parabolic trajectory, but its orbit was radically altered by a close approach to Jupiter into an elliptical orbit of 5.58 year period in 1767. It is likely another close encounter with Jupiter ejected it from the solar system. Asteroid 2010 JL33 is 99.2% probable of being a remnant of comet Lexells. The comet's size was 2.48-31 miles diameter, but most likely 18.6 miles.

Orbit formula—
$1 = x^2/3.14964^2 + y^2/1.948683^2$

perihelion= .6746 AU (8/14/1770 before Jupiter encounter)

aphelion= 2.363864 AU

a= 3.14964 AU

b= 1.948683 AU

eccentricity= .7856

1/a= 2.122452275 x 10^-12

period= 2 years, 54 days

Perihelion statistics—

Distance from sun at its closest= 73 million, 7 thousand, 373 miles

Velocity at closest distance from sun= 27 miles, 3,291 feet, 10.448 inches/second

Time of perihelion= 8/14/1770

Note—Now a lost comet.

Comet #23—

Name— Comet Kohoutek

Picture comet—

Picture discoverer —

Kohoutek

Type of comet — hyperbolic orbit

Comet Kohoutek was first seen on 3/7/1973 by Lubos Kohoutek. It reached perihelion on 12/28/1973 and was magnitude -3. It had a tail 25 degrees long with an antitail. Its last perihelion was 150,000 years ago and its next one will be in 75,000 years. This comet is an Oort Cloud comet. It was viewed from Skylab 4 and Soyuz 13.

Orbit formula— $1 = x^2/(-17,800^2) - y^2/71.2^2$

perihelion= .1424 AU (12/28/1973)

aphelion= n/a

a= -17,800 AU

b= 71.2 AU

eccentricity= 1.000008

period= n/a

Perihelion statistics—

Distance from sun at its closest= 13 million, 233 thousand, 515.66 miles

Velocity at closest distance from sun= 69 miles, 185 feet, 10.88 inches/second

Time of perihelion= 12/28/1973

Velocity at infinity=732 feet, 4.11 inches/ second

Note— This comet will not return to our solar system again.

Comet #24—
Name— Comet Swift Tuttle

Picture comet—

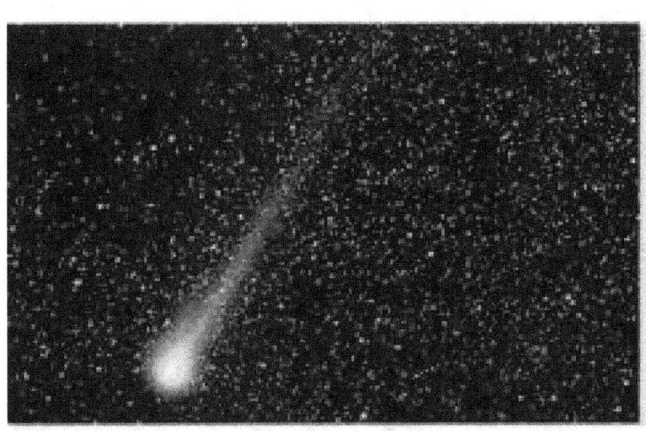

Picture discoverer—

Tuttle   Swift

Type of comet—elliptical orbit

Comet Swift Tuttle was discovered by Lewis Swift on 7/16/1862 and Horace Tuttle on 7/19/1862. It is 16.16 miles in diameter. The comet reached magnitude .1 in 188 AD as recorded by the Chinese, and was seen earlier in 69 BC and 322 BC. Also, it was seen 7/3/1737. On 8/14/2126, it will come periliously close to the earth or moon and will be magnitude .7. on 8/5/2126, it will come 14 million, 200 thousand miles near earth, 13 million, 700 thousand miles on 8/24/2261, 1 million miles in 3044, and on 9/15/4479, it has a one in a million chance

(.000002%) of hitting the earth when its distance from us will be 4 million, 650 thousand miles. If it does impact with earth, the energy released would be equivalent to 27 times the Cretaceous Paleocene impact.

Orbit formula—
$1 = x^2/26.092^2 + y^2/7.013157^2$

perihelion= .9632 AU
(12/11/1992, 7/2/2126)

aphelion= 51.225 AU
a= 26.092 AU
b= 7.013157 AU

eccentricity= .9632
1/a= 2.561929769 x 10^-13

period= 133 years, 102 days

Perihelion statistics—

Distance from sun at its closest= 89 million, 168 thousand, 246 miles

Velocity at closest distance from sun= 26 miles, 2,433 feet, 6.89 inches/second

Time of perihelion= 7/2/2126

Aphelion statistics—

Distance from sun at its farthest= 4 billion, 760 million, 441 thousand, 291 miles

Velocity at farthest distance from sun= 2,617 feet, 2.536 inches/second

Time of aphelion= 9/2/2059 AD

Returns— 7/2/2126

Comet #25—

Name— Great comet of 1843

Picture comet—

Picture discoverer— n/a

Type of comet—elliptical orbit

The Great Comet of 1843 was discovered on 2/5/1843. It is a member of the sun grazing Kreutz family of comets which resulted from the breakup of its parent comet x/1106-c1 into multiple fragments in 1106. It passed within a few solar radii of the sun and was very bright at magnitude -7 on 7/3/1843. It came to 77 million, 500 thousand miles of earth.

Orbit formula—
$1 = x^2/78^2 + y^2/.9228922^2$

perihelion = .00546 AU (2/27/1843)

aphelion = 156 AU

a = 78 AU

b = .9228922 AU

eccentricity = .99993

1/a = 8.56998353 x 10^-14

period = 688 years, 320 days

Perihelion statistics—

Distance from sun at its closest = 507 thousand, 409 miles

Velocity at closest distance from sun= 354 miles, 852 feet, 6.31 inches/second

Time of perihelion= 2/27/1843

Aphelion statistics—

Distance from sun at its farthest= 14 billion, 497 million, 390 thousand, 750 miles

Velocity at farthest distance from sun= .0098245 millimeters/second

Time of aphelion= 2187 AD

Returns— 2665 AD

Comet #26—
Name— Comet Wilson Harrington

Picture comet—

Picture discoverer—

Harrington   Helin

Type of comet—elliptical orbit

Comet Wilson Harrington was discovered by Robert Harrington and Albert Wilson on 11/15/1979. Eleanor Helin was a prolific discoverer of minor planets and several comets was the true discoverer of the comet. It is periodic comet and also an Apollo asteroid with a diameter of 2.45 miles, rotation period of 3 hours, 34 minutes and 25 seconds, an albedo of .05 and an absolute magnitude of 15.99.

Orbit formula—
$1 = x^2/2.6392^2 + y^2/2.05603^2$

perihelion= .98448 AU

aphelion= 4.2939 AU

a= 2.6392 AU

b= 2.05603 AU

eccentricity= .62698

1/a= 2.532868106 x 10^-12

period= 1,567 days

Perihelion statistics—

Distance from sun at its closest= 91 million, 489 thousand, 687.5 miles

Velocity at closest distance from sun= 23 miles, 4,165 feet, .5 inches/second

Time of perihelion= after 11/5/1979

Aphelion statistics—

Distance from sun at its farthest= 399 million, 40 thousand, 680.5 miles

Velocity at farthest distance from sun= 5 miles, 2,397 feet, .1689 inches/second

Time of aphelion= 12/28/1981

Returns— 2019 AD

Comet #27—

Name— Comet Wirtanen

Picture comet—

Picture discoverer—

Carl Wirtanen

Type of comet—elliptical orbit

Comet Wirtanin was discovered on 1/17/1948 by Carl Wirtanen. It is a small, short period Jupiter family comet, whose size is 3,937 feet in diameter, and has a rotation rate of 8 hours and 54 minutes. Its angular elongation was 20 degrees from the sun between January 23rd and September 26th of 2013. On 12/16/2018, the comet was 7 million, 190 thousand miles from the earth at magnitude 4.2. It was one of the closest comets to earth in 70 years. The Rosetta mission planned a visit to the comet, but instead went to comet 67p/Churyumov Gerasimeko.

Orbit formula—

$1 = x^2/3.094553^2 + y^2/2.331073^2$

perihelion= 1.0587602 AU
(12/12/2018, 5/19/2024)

aphelion= 5.129946 AU
a= 3.094553 AU
b= 2.331073 AU

eccentricity= .6576412

1/a= 2.160253652 x 10^-12

period= 1,986 days

Perihelion statistics—

Distance from sun at its closest= 98 million, 392 thousand, 713.5 miles

Velocity at closest distance from sun= 23 miles, 823 feet, 5.935 inches/second

Time of perihelion= 5/19/2024

Aphelion statistics—

Distance from sun at its farthest= 476 million, 736 thousand, 100 miles

Velocity at farthest distance from sun= 4 miles, 4,113 feet, 8.831 inches/second

Time of aphelion= 9/2/2021

Returns— 5/19/2024

Comet #28—
Name— Comet Pons Winnicke

Picture comet—

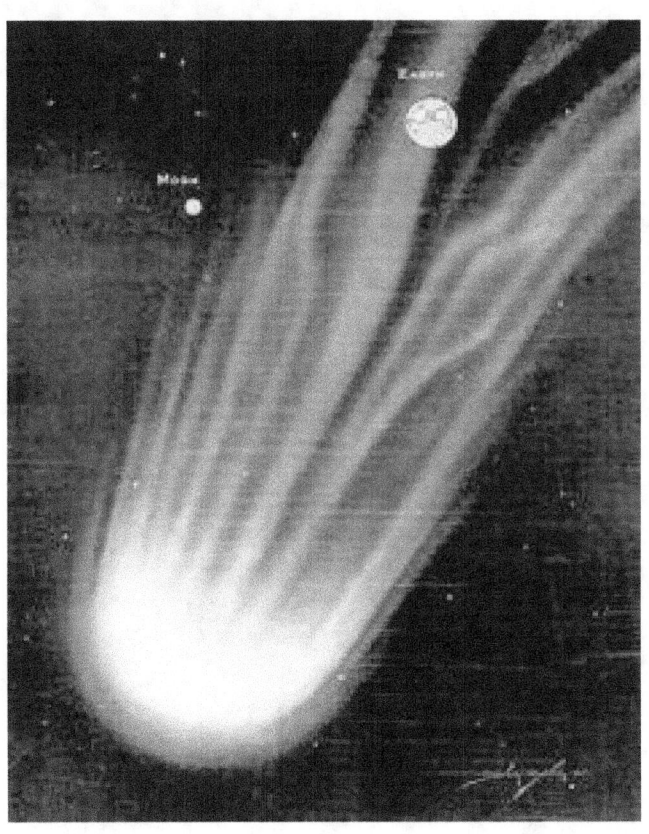

Picture discoverer—

Pons

Type of comet—elliptical orbit

Comet Pons Winnicke is a periodic Jupiter family comet discovered by Jean Louis Pons and Friedrich Winnicke on

6/12/1919 and 3/9/1858, respectively. Its most recent perihelion was on 1/30/2015. The comet produced impressive meteor showers in 1916, 1921, and 1927. It came to a distance of 3 million, 720 thousand miles of earth in June 1927. There will be a close approach to Jupiter in July 2037 and change the perihelion to .982 AU. The comet is 3 miles, 1,183 feet in diameter.

Orbit formula—

$1 = x^2/3.434^2 + y^2/2.6556^2$

perihelion = 1.257 AU
(9/26/2008, 5/27/2021)

aphelion = 5.611 AU

a = = 3.434 AU

b = 2.6556 AU

eccentricity = .634

$1/a = 1.946589154 \times 10^{-12}$

period = 2,309 days

Perihelion statistics—

Distance from sun at its closest= 116 million, 815 thousand, 514 miles

Velocity at closest distance from sun= 9 miles, 5,206 feet, 4.15 inches/second

Time of perihelion= 9/26/2008

Aphelion statistics—

Distance from sun at its farthest= 521 million, 441 thousand, 407 miles

Velocity at farthest distance from sun= 4 miles, 3,835 feet, 9.608 inches/second

Time of aphelion= 11/24/2011

Returns— 2021 AD

Comet #29—

Name— Comet Hartley

Picture comet—

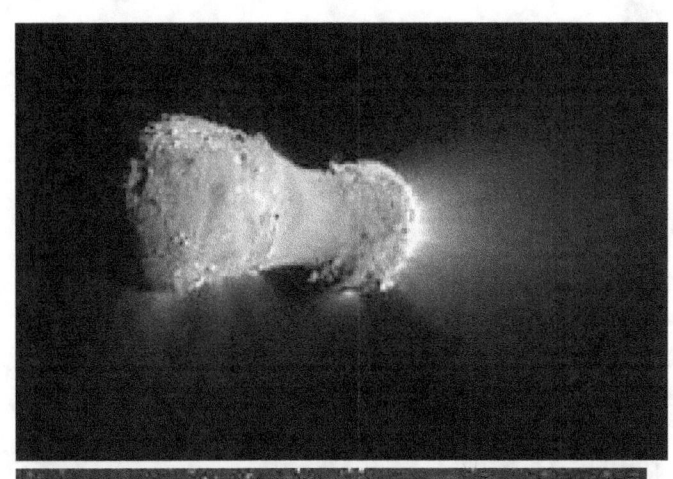

Picture discoverer—

Hartley

Type of comet—elliptical orbit

Comet Hartley is a small, peanut shaped, periodic comet discovered by Malcolm Hartley on 3/15/1996. On 10/20/2010, 7 days before perihelion, the comet was 11 million, 160 thousand miles from earth. It is 3,960 x 5,227 feet in dimensions, weighs 330 million, 750 thousand tons, has an albedo of .028, and should survive for another 100 orbits (approximately 700 years). The surface of the comet has blocky objects 165 feet high and 260 feet wide (size of a 16 story building). Deep Impact mission spacecraft visited the comet 11/4/2010 and came to within 435 miles of it while moving at

27,500 mph. The comet was the second one visited by the spacecraft; the first was Temple 1.

Orbit formula—
$1 = x^2/3.46^2 + y^2/2.49^2$

perihelion = 1.05 AU (4/20/2017, 10/12/2023)

aphelion = 5.87 AU

a = 3.46 AU

b = 2.49 AU

eccentricity = .694

$1/a = 1.91535448 \times 10^{-12}$

period= 2,358 days

Perihelion statistics—

Distance from sun at its closest= 97 million, 560 thousand, 999 miles

Velocity at closest distance from sun= 23 miles, 2,862 feet, 8 inches/second

Time of perihelion= 10/12/2023

Aphelion statistics—

Distance from sun at its farthest= 545 million, 510 thousand, 793 miles

Velocity at farthest distance from sun= 4 miles, 452 feet, 2.32 inches/second

Time of aphelion= 7/3/2020

Returns— 10/12/2023

Comet #30—
Name— Comet Gehrels

Picture comet—

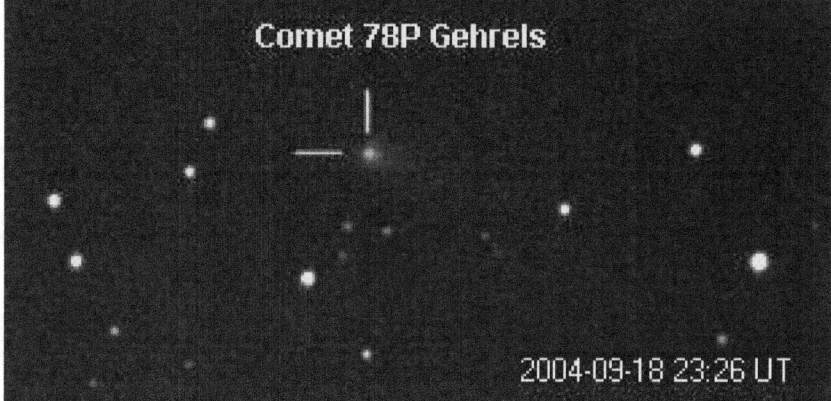

Picture discoverer—

Tom Gehrels

Type of comet—elliptical orbit

Comet Gehrels was discovered by Tom Gehrels on 9/29/1973 at magnitude 15. It is a periodic, Jupiter family comet. In 1989 and 1997, under favorable conditions, it increased to magnitude 12, and was later observed to reach magnitude 10 in 2012 and 2018. On 9/15/2029, it will be 1 million, 674 thousand miles from Jupiter and the comet's orbit will be strongly perturbed and changed.

Orbit formula—

$1 = x^2/3.735^2 + y^2/3.31211^2$

perihelion= 2.009 AU (1/12/2012, 4/2/2019)

aphelion= 5.462 AU

a= 3.735 AU

b= 3.31211 AU

eccentricity= .4622

$1/a = 1.789715436 \times 10^{-12}$

period = 2,635 days

Perihelion statistics—

Distance from sun at its closest = 186 million, 700 thousand, 372 miles

Velocity at closest distance from sun = 15 miles, 4,153 feet, 8.593 inches/second

Time of perihelion = 4/2/2019

Aphelion statistics —

Distance from sun at its farthest = 507 million, 594 thousand, 540.4 miles

Velocity at farthest distance from sun = 5 miles, 4,253 feet, 10.05 inches/second

Time of aphelion = 11/10/2022

Returns — 2026 AD

Comet #31 —

Name — Comet ISON

Picture comet —

Picture discoverer —

Nevsky

Type of comet—hyperbolic orbit

Comet ISON is a hyperbolic, sun grazing comet discovered on 9/21/2012 by Vitaly Nevsky and A. Novichonok. The comet is freshly from the Oort Cloud and is .5 miles in diameter. It completely broke up hours before it past close to the sun on 11/28/2013 when it was 1 million, 162 thousand, 500 miles from the center of the sun, just 724,000 from the surface of the sun. It reached magnitude -3 to -5.

Orbit formula— $1 = x^2/(-62.2^2) - y^2/1.244^2$

perihelion= .01244 AU (11/28/2013)

aphelion= n/a
a= -62.2 AU
b= 1.244 AU

eccentricity= 1.0002

period= n/a

Perihelion statistics—

Distance from sun at its closest= 1 million, 155 thousand, 679.5 miles

Velocity at closest distance from sun= 165 miles, 4,819 feet, 10.87 inches/ second

Time of perihelion= 11/28/2013

Velocity at infinity=2 miles, 1,828 feet, 9.5 inches/second

Note—This comet has disintegrated.

Comet #32—

Name— The Great Comet of 1910

Picture comet—

Picture discoverer—

Innes

Type of comet—elliptical orbit

The Great Comet of 1910 was first studied by Robert Innes, a Scottish astronomer. It was a daylight comet which outshone the planet Venus and was possibly the brightest comet in the 20th century. By January 2, 1910, it was at magnitude -1. At perihelion on January 17, 1910, it was at magnitude -5. Early in February, it had a curved tail 50 degrees long.

Orbit formula—
$1 = x^2/1487^2 + y^2/4.70234^2$

perihelion= .128975 AU (1/17/1910)

aphelion= 2,973.000005 AU

a= 1,487 AU

b= 4.70234 AU

eccentricity= .999995

1/a= 4.495351146 x 10^-15

period= 57,300 years

Perihelion statistics—

Distance from sun at its closest= 11 million, 985 thousand, 903.67 miles

Velocity at closest distance from sun= 72 miles, 4,588 feet, 6.28 inches/second

Time of perihelion= 1/17/1910

Aphelion statistics—

Distance from sun at its farthest= 276 billion, 286 million, 812 thousand, 700 miles

Velocity at farthest distance from sun= 14.164 millimeters/second

Time of aphelion= 30,560 AD

Returns— 59,210 AD

Comet #33—

Name— Comet Giacobinai Zinner

Picture comet—

Picture discoverer— n/a

Type of comet—elliptical orbit

Comet Giacobinai Zinner was discovered by Michel Giacobinai in the constellation of Aquarius on 12/20/1900. It is 1 mile, 1,282 feet in diameter, and is responsible for the Draconid meteor shower. The comet reached magnitude 7 or 8 during its appearance. On 9/11/1985, the space probe International Cometary Explorer passed through the comet's tail.

Orbit formula—

$1 = x^2/3.526^2 + y^2/2.49856^2$

perihelion = 1.038 AU (9/10/2018, 3/25/2025)

aphelion = 6.014 AU

a = 3.526 AU

b = 2.49856 AU

eccentricity = .7056

1/a = 1.89578966 x 10^-12

period = 2,417 days

Perihelion statistics—

Distance from sun at its closest= 96 million, 556 thousand, 340 miles

Velocity at closest distance from sun= 23 miles, 3,806 feet, 7.496 inches/second

Time of perihelion= 9/10/2018

Aphelion statistics—

Distance from sun at its farthest= 558 million, 893 thousand miles

Velocity at farthest distance from sun = 4 miles, 497 feet, 2.71 inches/second

Time of aphelion = 1/31/2022

Returns — 3/25/2025

Comet #34 —
Name — Comet Stephan Oterma

Picture comet—

Picture discoverer— n/a

Type of comet—elliptical orbit

Comet Stephan Oterma is a Halley type comet (period 20-200 years) which was discovered by Jerome Eugene loggia in January 1867. It is considered a Centaur since its perihelion is near mar's orbit, and its aphelion is nears Uranus' orbit. The comet has no coma.

Orbit formula—
$1 = x^2/11.247^2 + y^2/5.738898^2$

perihelion= 1.5744 AU (11/10/2018, 8/28/2056)

aphelion= 20.92 AU a= 11.247 AU

b= 5.738898 AU

eccentricity= .86002

1/a= 5.943440165 x 10^-13

period= 37 years, 263 days

Perihelion statistics—

Distance from sun at its closest= 146 million, 312 thousand, 128 miles

Velocity at closest distance from sun= 20 miles, 600 feet/second

Time of perihelion= 11/10/2018

Aphelion statistics—

Distance from sun at its farthest= 1 billion, 944 million, 137 thousand, 273 miles

Velocity at farthest distance from sun= 1 miles, 2,711 feet, 5.784 inches/second

Time of aphelion= 8/21/2037

Returns— 8/28/2056

Comet #35—
Name— Comet d'Arrest

Picture comet—

Picture discoverer—

d'Arrest

Type of comet—elliptical orbit

Comet d'Arrest was first observed by Heinrich Ludwig d'Arrest on 28-30 of June, 1851. It is a periodic, Jupiter family comet which orbits between Mars and Jupiter. On 8/12/1976, it cane within 14 million, 59 thousand miles of earth. The 30 degrees elongation of its tail occurred during its return from October 2014 to May 2015. It is 1 mile, 5,196 feet in diameter.

Orbit formula—
$1 = x^2/3.495^2 + y^2.7622^2$

perihelion= 1.353 AU (3/2/2015, 9/17/2021)

aphelion= 5.637 AU

a= 3.495 AU

b= 2.7622 AU

eccentricity= .6127

1/a= 1.91261435 x 10^-12

period= 6 years, 197 days

Perihelion statistics—

Distance from sun at its closest= 125 million, 736 thousand, 985 miles

Velocity at closest distance from sun= 20 miles, 1,078 feet, 2.4 inches/second

Time of perihelion= 3/2/2015

Aphelion statistics—

Distance from sun at its farthest= 523 million, 857 thousand, 639 miles

Velocity at farthest distance from sun= 4 miles, 4,485 feet, 11.938 millimeters/ second

Time of aphelion= 6/9/2018

Returns— 9/17/2021

Comet #36—

Name— Comet Tempel Tuttle

Picture comet—

Picture discoverer—

Tempel   Tuttle

Type of comet—elliptical orbit

Comet Temple Tuttle was discovered by Wilhelm Temple and Horace Tuttle on

12/19/1965 and 1/6/1861, respectively. It has a size of 2 miles, 1,251 feet diameter, has a mass of 13 billion, 23 million tons, and has a mass stream of 5 billion, 125 million tons. The comet of 1366 was the comet Tempel Tuttle on a previous return, and it was 2 million, 130 thousand miles from earth then. This comet is responsible for the Leonids meteor shower. There is a 33 year cycle for when these showers occur. In November 2009, the earth passed through the meteor shower which was the result of the comet's 1466 and 1533 appearances.

Orbit formula—

$1 = x^2/10.3345^2 + y^2/4.3854^2$

perihelion= .9766 AU (2/28/1998, 5/20/2031)

aphelion= 19.6924 AU a= 10.3345 AU

b= 4.3854 AU

eccentricity= .9055

$1/a = 6.468225026 \times 10^{-13}$

period = 33 years, 81 days

Perihelion statistics—

Distance from sun at its closest = 90 million, 757 thousand, 383 miles

Velocity at closest distance from sun = 25 miles, 4,480 feet, 5.15 inches/second

Time of perihelion = 2/28/1998

Aphelion statistics—

Distance from sun at its farthest= 1 billion, 830 million, 53 thousand, 960 miles

Velocity at farthest distance from sun= 1 miles, 1,488 feet, 5.255 inches/second

Time of aphelion= 12/20/2013

Returns— 5/20/2031

Comet #37 —
Name — Comet Ikeya Zhang
Picture comet —

Picture discoverer—

Ikeya　　　　Zhang

Type of comet—elliptical orbit

Comet Ikeya Zhang was discovered on 2/1/2002 by

Kaoru Ikeya and Zhang Daqing in the constellation of Cetus. It has the longest orbiting period of a periodic comet, 366 years and 186 days. It reached perihelion on 3/18/2002 and was apparent magnitude 3.5 in brightness, being the most luminous comet since 1997. It has a diameter of 12 miles, 2,257 feet. Johannes Hevelius, a Polish astronomer, observed this comet in 1661.

Orbit formula—

$1 = x^2/51.2136^2 + y^2/7.18926^2$

perihelion= .507141 AU (3/18/2002, 9/1/2362)

aphelion= 101.92 AU

a= 51.2136 AU

b= 7.18926 AU

eccentricity= .990098

1/a= $1.305236725 \times 10^{-13}$

period= 366 years, 186 days

Perihelion statistics—

Distance from sun at its closest= 47 million, 129 thousand, 623 miles

Velocity at closest distance from sun= 36 miles, 3,471 feet, 8.286 inches/second

Time of perihelion= 3/18/2002

Aphelion statistics—

Distance from sun at its farthest= 9 billion, 471 million, 628 thousand, 627 miles

Velocity at farthest distance from sun= 963 feet, 1.733 inches/second

Time of aphelion= 6/19/2185

Returns— 9/1/2362

Comet #38—

Name— Comet Thatcher

Picture comet—

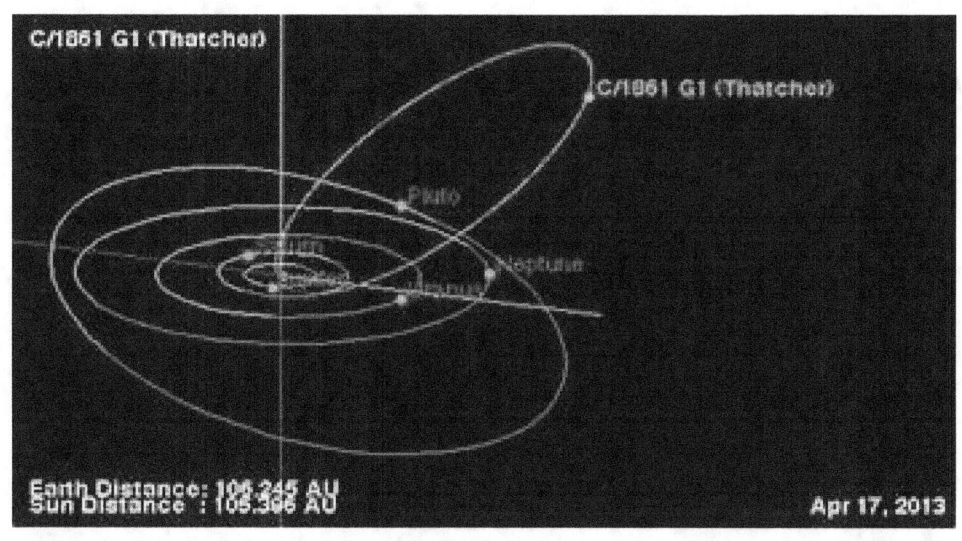

Picture discoverer — n/a

Type of comet — elliptical orbit

Comet Thatcher is a long period comet discovered by A.E. Thatcher. It was closest to the earth on 5/5/1861 at 31 million, 155 thousand miles, and came to perihelion on 6/3/1861. It produces the Lyrid meteor shower.

Orbit formula—

$1 = x^2/55.6^2 + y^2/10.20847^2$

perihelion= .9207 AU
(6/3/1861, 2280+/-5 AD)

aphelion= 110 AU

a= 55.6 AU

b= 10.20847 AU

eccentricity= .983

1/a= 1.202263877 x 10^-13

period= 415 years

Perihelion statistics—

Distance from sun at its closest= 85 million, 562 thousand, 485 miles

Velocity at closest distance from sun= 27 miles, 848 feet, 5.165 inches/second

Time of perihelion= 9/10/2018

Aphelion statistics—

Distance from sun at its farthest= 10 billion, 222 million, 519 thousand, 120 miles

Velocity at farthest distance from sun= 1,368 feet, 7.37 inches/second Time of aphelion= 2069 AD

Returns— 2280 AD

Comet #39—
Name— Comet 2010 x1 Elenin

Picture comet—

Picture discoverer —

Leonid Elenin

Type of comet — hyperbolic orbit

Comet Elenin is an Oort Cloud comet discovered by Leonid Elenin on 12/10/2010. It has a diameter of 3,937 feet. In August 2011, the coma was 122,000 miles in size. August

19, 2011, a coronal mass ejection hit the comet and caused it to start to disintegrate which caused a debris field like the comet Shoemaker Levy did on June 23, 1993.

It reached magnitude 6 by September and October 2011. The heliocentric orbital period was 600,000 years. Before it entered the solar system, it was tens of millions of years, and its aphelion was 105,691 (1.67 light years). This comet is probably an outer Oort Cloud comet that was easily perturbed by a passing star.

Orbit formula— $1 = x^2/(-518^2) - y^2/5.79294^2$

perihelion= .48243 AU (9/10/2011)

aphelion= 1,037 AU

a= -518 AU

b= 5.79294 AU

eccentricity= 1.0000621

period= n/a (disintegrated)

Perihelion statistics—

Distance from sun at its closest= 44 million, 883 thousand, 181 miles

Velocity at closest distance from sun= 37 miles, 3,533 feet, 7.441 inches/second

Time of perihelion= 9/10/2011

Velocity at infinity=4,293 feet, 1.77 inches/second (Would have been this figure if the comet had not disintegrated.)

Comet #40—

Name— Comet Skjellerup Maristany

Picture comet—

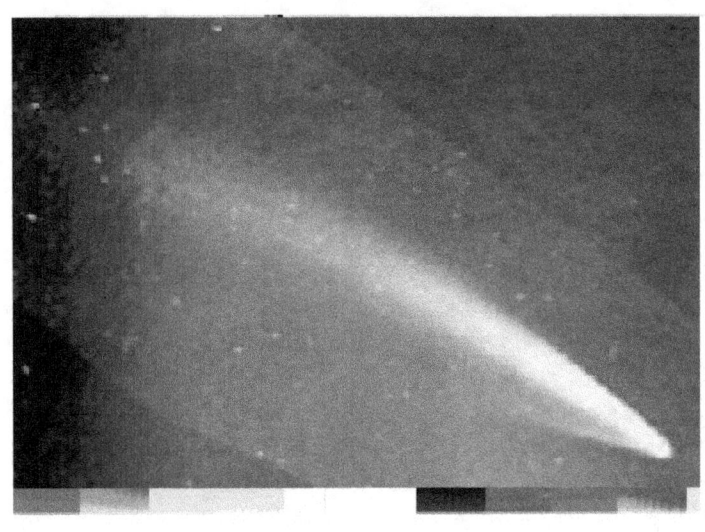

picture of discoverer— n/a

type of comet— elliptical orbit

Comet Skjellerup Maristany is a very bright, long period comet discovered on 12/6/1927 by John Francis Skjellerup. It could be observed with the naked eye for 32 days at peak magnitude of -6 on 12/6/1927, and could be seen if one blocked the sun 1.4 degrees from the sun on 12/15/1927. It came within 68 million, 200 thousand miles of earth. As of 2010, it is more than 9 billion, 765 million miles from the sun.

Orbit formula—

$1 = x^2/1{,}101^2 + y^2/22.02^2$

perihelion= .1761 AU (12/18/1927)

aphelion= 2,201.5596 AU

a= 1,101 AU

b= 22.02 AU

eccentricity= .9998

1/a= 1.89578966 x 10^-12

period= 36,532 years 231 days

Perihelion statistics—

Distance from sun at its closest= 16 million, 365 thousand, 324 miles

Velocity at closest distance from sun= 62 miles, 1,902 feet, 4.467 inches/second

Time of perihelion= 12/18/1927

Aphelion statistics—

Distance from sun at its farthest= 204 billion, 595 million, 319 thousand, 200 miles

Velocity at farthest distance from sun = 41 feet, 7.7905 inches/second

Time of aphelion = 20,193 AD

Return time — 38,460 AD

Comet #41 —
Name — Comet Humason

Picture comet — not found

Picture discoverer—

Humason

Type of comet—elliptical orbit

Comet Humason was discovered by Milton Humason on 9/2/1962. It is a giant comet 18.6 x 25.42 miles in dimensions, and was magnitude 1.5 to 3.5.

Orbit formula—
$1 = x^2/204.52^2 + y^2/36.9977^2$

perihelion= 2.1334 AU (12/10/1962)

aphelion= 406.91 AU
a= 204.52 AU
b= 36.997 AU

eccentricity= .98956

$1/a = 3.268399956 \times 10^{-14}$

period= 2,940 days

Perihelion statistics—

Distance from sun at its closest= 198 million, 261 thousand, 112 miles

Velocity at closest distance from sun= 17 miles, 459 feet, 6.7 inches/second

Time of perihelion= 12/10/1962

Aphelion statistics—

Distance from sun at its farthest= 37 billion, 814 million, 956 thousand, 870 miles

Velocity at farthest distance from sun = 796 feet, 4.15 inches/second

Time of aphelion = 2026 AD

Return time — 2022 AD

Comet #42 —
Name — Comet Bennet

Picture comet —

Picture discoverer— n/a
Type of comet—elliptical orbit

Comet Bennet was discovered by John Caister Bennet on 12/28/1969. It is one of two bright comets in the 1970s, the other being Comet West. It reached perihelion on 3/20/1970 and was closest to earth on the 26th of march. Its peak magnitude was 0, and it was last seen on 2/27/1971.

Orbit formula—
$1 = x^2/141^2 + y^2/12.2804^2$

perihelion= .538 AU (3/20/1970)

aphelion= 281.462 AU
a= 141 AU
b= 12.2804 AU

eccentricity= .9962

1/a=4.740841953 x 10^-14

period= 1,678 years

Perihelion statistics—

Distance from sun at its closest= 49 million, 997 thousand, 412 miles

Velocity at closest distance from sun= 35 miles, 3,405 feet, 11.35 inches/second

Time of perihelion= 3/20/1970

Aphelion statistics—

Distance from sun at its farthest= 26 billion, 824 million, 340 thousand miles

Velocity at farthest distance from sun= 359 feet, 8.95 inches/second

Time of aphelion= 2809 AD

Return time— 3648 AD

Comet #43—
Name— Comet Mcnaught

Picture comet—

Picture discoverer—

Robert Mcnaught

Type of comet—elliptical orbit

Comet Mcnaught was the great comet of 2007. It was discovered by Robert Mcnaught on 8/7/2006 by British astronomer Robert Mcnaught. It was the brightest comet seen in over 40 years since comet Ikeya

Seki, and was easily visible to the naked eye at magnitude -5.5. It reached perihelion on 1/12/2007 and its tail had a length of 35 degrees. It was closest to earth on 1/15/2007 at 72 million, 260 thousand miles. The spacecraft Ulysses passed through the comet's ion trail on 2/3/2007. The comet once had a hyperbolic orbit as it entered the solar system from the Oort Cloud, but its eccentricity is now less than 1.

Orbit formula—

$1 = x^2/2050^2 + y^2/108.9218^2$

perihelion = .170754 AU (1/12/2007)

aphelion = 69,481.14937 AU

a = 2050 AU

b = 108.9218 AU

eccentricity = was 1.000019, but now .999995084

$1/a = 1.910610776 \times 10^{-16}$

period = incoming was 650,000 years, but now 92,817 years, 255 days

Perihelion statistics—

Distance from sun at its closest= 15 million, 868 thousand, 509 miles

Velocity at closest distance from sun= 132 miles, 3,184 feet, 2.09 inches/second

Time of perihelion= 1/12/2007

Aphelion statistics—

Distance from sun at its farthest= 6 trillion, 458 billion, 569 million, 491 thousand miles

Velocity at farthest distance from sun= 3,006 feet, 9.46 inches/second

Time of aphelion= 48,416 AD

Return date=94,824 AD

Comet #44—
Name— Comet Morehouse

Picture comet—

Picture discoverer—

DANIEL W. MOREHOUSE

Daniel Walter Morehouse

Type of comet—hyperbolic orbit

Comet Morehouse was a bright, non periodic comet discovered by Daniel Walter Morehouse on

9/1/1908. Its tail had rapid variations and split into 6 separate tails, and at another time, the tail detached from its nucleus. The tail formed 186,000,000 miles from the sun while they usually form about 139,500,000 miles from the sun. This comet is a typical one fresh from the Oort cloud.

Orbit formula— $1 = x^2/(-1,295.5055^2) - y^2/49.51^2$

perihelion= .945 AU (1/12/2007)

aphelion= n/a

a= -1,295.5055 AU

b= 49.51 AU

eccentricity= 1.00073

period= n/a (Has no period and will not return.)

Perihelion statistics—

Distance from sun at its closest= 87 million, 820 thousand, 732 miles

Velocity at closest distance from sun= 26 miles, 4,832 feet, 1.85 inches/second

Time of perihelion= 1908

Velocity at infinity= 4,369 feet, 9.98 inches/second

Note—This comet will not return.

Comet #45—
Name— Comet Grigg Skjullerup

Picture comet—

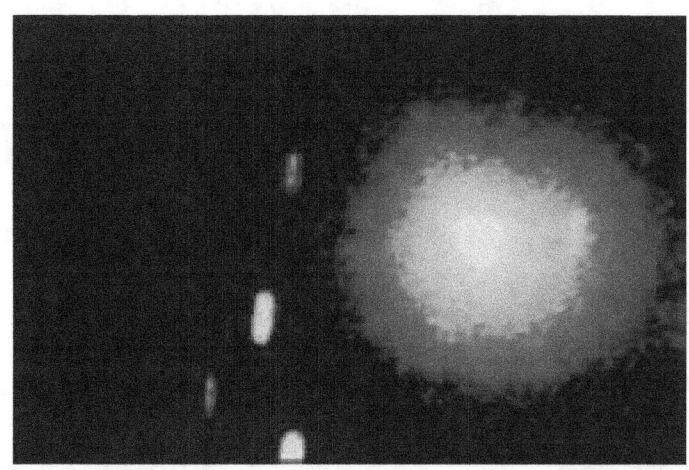

Picture discoverer— n/a

Type of comet—elliptical orbit

Comet Grigg Skjellerup is a periodic comet visited by the Giotto spacecraft in July 1992 after it first visited Halley's comet. It was discovered by John Skjellerup and John Grigg on 7/23/1902. Jupiter's gravitational influence altered its orbit considerably, making the perihelion .77 in 1725, .89 in 1922, .99 in 1977, and 1.12 in 1999. It is responsible for the Pi Puppids meteor shower of 1972 which peaked April 23rd in the Southern Hemisphere. The comet's size is 1 mile, 3,231 feet in diameter, and the mineral Brownlite comprises the comet.

Orbit formula—
$1 = x^2/3.0437^2 + y^2/2.6918^2$

perihelion= 1.1168 AU
(3/23/2008, 12/25/2023)

aphelion= 4.9332 AU

a= 3.0437 AU

b= 2.6918 AU

eccentricity= .6631

1/a= 2.196204341 x 10^-12

period= 1,938 days

Perihelion statistics—

Distance from sun at its closest= 103 million, 786 thousand, 449 miles

Velocity at closest distance from sun= 22 miles, 2,192 feet, 10.82 inches/second

Time of perihelion= 3/23/2008

Aphelion statistics—

Distance from sun at its farthest= 458 million, 452 thousand, 103 miles

Velocity at farthest distance from sun= 5 miles, 689 feet, 4.81 inches/second

Time of aphelion= 11/17/2010

Return time= 12/25/2023

Comet #46—
Name— Comet Tempel 2
Picture comet—

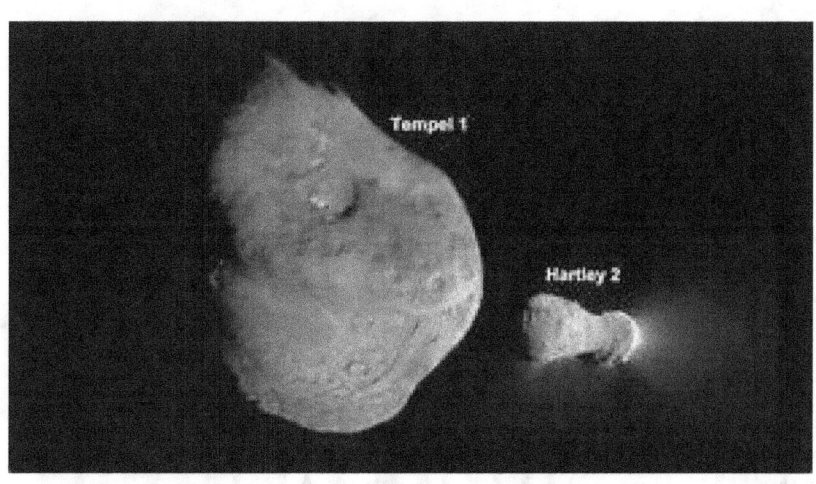

Picture discoverer—

Tempel

Type of comet—elliptical orbit

Comet Tempel 2 is a bright periodic comet of magnitude 7 and has an orbit of 5 years, 146 days. It was discovered by Wilhelm Tempel about 1873.

Orbit formula—

$1 = x^2/3.04^2 + y^2/2.1989^2$

perihelion= 1.38 AU (1873)

aphelion= 4.7 AU

a= 3.04 AU

b= 2.1989 AU

eccentricity= .55

1/a= 2.198877353 x 10^-12

period= 5 years, 146 days

Perihelion statistics—

Distance from sun at its closest= 128 million, 246 thousand, 149 miles

Velocity at closest distance from sun= 19 miles, 3,098 feet, 1.74 inches/second

Time of perihelion= 1873

Aphelion statistics—

Distance from sun at its farthest= 436 million, 780 thousand, 362.5 miles

Velocity at farthest distance from sun= 5 miles, 3,965 feet, 3.93 inches/second

Time of aphelion= 2022

Return time— 2019 AD

Comet #47—

Name— Comet Schwassman Wachmann 2

Picture comet—

Picture discoverer—

Friedrich Schwassman

Type of comet—elliptical orbit

Comet Schwassmann Wachmann 2 is a periodic comet discovered by Friedrich Schwassman and Arno Arthur

Wachmann on 1/17/1929. It is 3 miles, 4,456 feet in diameter and was magnitude 11 when discovered. It is seen at every one of its returns.

Orbit formula—

$1 = x^2/3.485^2 + y^2/3.20904^2$

perihelion= 2.14

aphelion= 4.83 AU

a= 3.485 AU

b= 3.20904 AU

eccentricity= .39

1/a= $1.918102483 \times 10^{-12}$

period= 6 years, 183 days

Perihelion statistics—

Distance from sun at its closest= 198 million, 874 thousand, 463 miles

Velocity at closest distance from sun= 14 miles, 4,710 feet, 2.52 inches/second

Time of perihelion= 1929

Aphelion statistics—

Distance from sun at its farthest= 448 million, 861 thousand, 521 miles

Velocity at farthest distance from sun= 6 miles, 3,158 feet, 2.76 inches/second

Time of aphelion= 2027

Return time= 2024/2025

Comet #48—

Name— Comet Brooks 2

Picture comet—

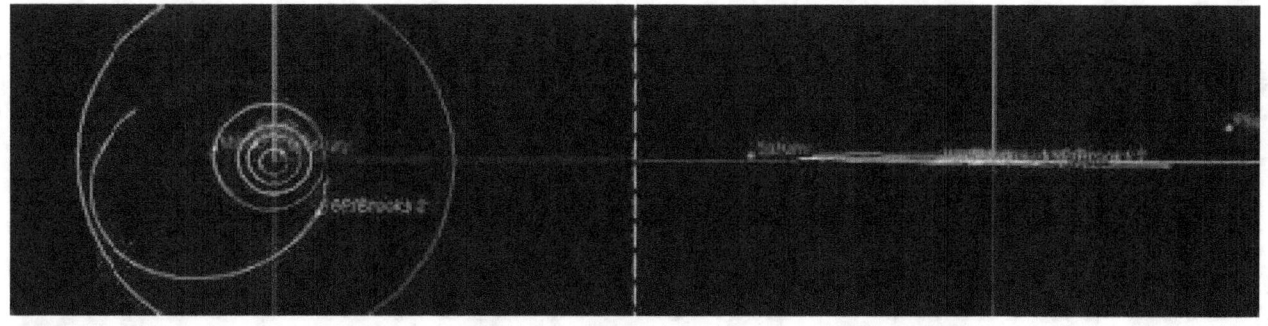

Picture discoverer— n/a

Type of comet—elliptical orbit

Comet Brooks 2 is a periodic comet discovered by William Robert Brooks on 7/7/1889 in the constellation of Aquarius. In 1886, it passed the Roche limit of Jupiter, spent 2 days in orbit around the moon Io, and the comet was shattered; the fragments were seen in 1889. This encounter with Jupiter lowered the perihelion from 1.96 to 1.86. On 12/31/2016, the comet was 30 million, 969 thousand miles from Jupiter, and will be 22 million, 971 thousand miles from the planet on 7/3/2053.

Orbit formula—

$1 = x^2/3.63^2 + y^2/3.16435^2$

perihelion= 1.85 (1889)

aphelion= 5.41 AU

a= 3.63 AU

b= 3.16435 AU

eccentricity= .49

$1/a$ = 1.841484064 x 10^{-12}

period= 6 years, 219 days

Perihelion statistics—

Distance from sun at its closest= 171 million, 924 thousand, 185 miles

Velocity at closest distance from sun= 16 miles, 3.216 feet, 9.26 inches/second

Time of perihelion= 1889

Aphelion statistics—

Distance from sun at its farthest= 495 million, 404 thousand, 508 miles

Velocity at farthest distance from sun= 5 miles, 3,588 feet, 6.645 inches/second

Time of aphelion= 2022

Return time= 2024

Comet #49 —

Name — Comet Finley

Picture comet —

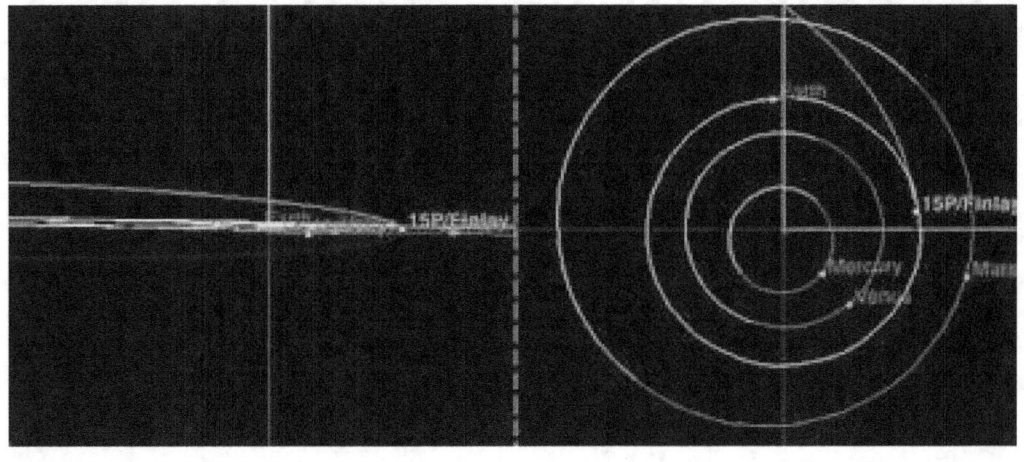

Picture discoverer — n/a

Type of comet — elliptical orbit

Comet Finley was discovered by William Henry Finley on 9/26/1886, and came to perihelion recently on 12/27/2014. Since 1953, it has been seen at its every return. On 10/25/2060, it will be 3 million, 700 thousand miles from earth. It is 1 mile, 612 feet in diameter, but this figure is uncertain.

Orbit formula—

$1 = x^2/3.645^2 + y^2/2.60305^2$

perihelion = .1.1 (1886)

aphelion = 6.19 AU

a = 3.645 AU

b = 2.60305 AU

eccentricity = .7

1/a = 1.833905945 x 10^-12

period = 7 years

Perihelion statistics—

Distance from sun at its closest= 102 million, 225 thousand, 191 miles

Velocity at closest distance from sun= 22 miles, 5,241 feet, 7.05 inches/second

Time of perihelion= 2019

Aphelion statistics—

Distance from sun at its farthest= 575 million, 294 thousand, 31 miles

Velocity at farthest distance from sun= 4 miles, 453 feet, 9.45 inches/second

Time of aphelion= 2021

Return time= 2024

Comet #50—

Name— Comet Faye

Picture comet— n/a

Picture discoverer—

Herve Faye

Type of comet—elliptical orbit

Comet Faye was discovered on 11/23/1843 by Herve Faye. It is a Jupiter family comet and is 2 miles, 898 feet in diameter. It made a return on 5/9/2006 and was visible to the naked eye till the end of November.

Orbit formula—

$1 = x^2/3.775^2 + y^2/1.77075^2$

perihelion= 1.59 AU (1843)

aphelion= 5.96 AU

a= 3.775 AU

b= 1.77075 AU

eccentricity= .58

1/a= $1.770751564 \times 10^{-12}$

period= 4 years, 146 days

Perihelion statistics—

Distance from sun at its closest= 147 million, 761 thousand, 867 miles

Velocity at closest distance from sun= 18 miles, 2,322 feet, 2.025 inches/second

Time of perihelion= 1843

Aphelion statistics—

Distance from sun at its farthest= 553 million, 874 thousand, 672 miles

Velocity at farthest distance from sun= 4 miles, 4,854 feet, 1.63 inches/second

Time of aphelion= 2022/2023

Return time= 2026

Comet #51—
Name— Comet Wolf 1
Picture comet— n/a
Picture discoverer— n/a
Type of comet—elliptical orbit

Comet Wolf 1 was discovered 9/17/1854 by Max Wolf. It had a previous perihelion of 2.74

with a period of 8 years, 234 days, but now the perihelion is 2.43 and period 8 years, 102 days. The current parameters are due to a passing of Jupiter at a distance of 11 million, 625 thousand miles on 9/27/1922, and the parameters are still the later set as of 8/13/2005. Another Jupiter approach on 3/10/2041 could return the parameters to their original ones.

Orbit formula—

$1 = x^2/4.075^2 + y^2/1.6404^2$

perihelion = 2.42

aphelion = 5.73 AU

a = 4.075 AU

b = 1.6404 AU

eccentricity = .4

1/a = 1.640389486 x 10^-12

period = incoming was 650,000 years, but now 92,817 years, 255 days

Perihelion statistics—

Distance from sun at its closest= 224 million, 895 thousand, 421 miles

Velocity at closest distance from sun= 14 miles, 558 feet, 4.044 inches/second

Time of perihelion= 1894

Aphelion statistics—

Distance from sun at its farthest= 857 million, 195 thousand, 799 miles

Velocity at farthest distance from sun= 5 miles, 5,055 feet, .89 inches/second

Time of aphelion= 48,263 AD

Return time= 94,671 AD

Comet #52—

Name— Comet Tuttle

Picture comet—

Picture discoverer—

Tuttle

Type of comet—elliptical orbit

Comet Tuttle is a Jupiter family class comet. On 12/30/2007, if was in close conjunction with galaxy M33. In January 2008, it was in the constellation of Eridanus and its distance from earth was 23 million, 512 thousand 260 miles from earth. The comet is responsible for the

Ursid meteor shower in late December. The comet is also 2 miles, 4,171 feet in diameter.

Orbit formula—
$1 = x^2/5.73^2 + y^2/3.279643^2$

perihelion= 1.01 AU

aphelion= 10.45 AU

a= 5.73 AU

b= 3.279643 AU

eccentricity= ,82

1/a= 1.166994617 x 10^-12

period= 13 years. 256 days

Perihelion statistics—

Distance from sun at its closest= 93 million, 861 thousand, 312 miles

Velocity at closest distance from sun= 24 miles, 4,573 feet, 9.986 inches/second

Time of perihelion= 1790

Aphelion statistics—

Distance from sun at its farthest = 971 million, 139 thousand, 317 miles

Velocity at farthest distance from sun = 2 miles, 2,129 feet, 7.71 inches/second

Time of aphelion = 2030

Return time = 2023

Comet #53 —

Name — Great Comet Donati

Picture comet —

Picture discoverer—

Giovanni Battista Donati

Type of comet—elliptical orbit

Comet Donati was, after the 1811 comet, the most brilliant of the 19th century. It was discovered by Giovanni Battista Donati on 6/2/1858 and the first to be photographed. On 5/7/1858, it was near the head of the constellation of Leo. By mid August, it was a naked eye object. In September, it was in Ursae Major, then nearest to the earth of 10/10/1858 at 49 million, 697 thousand miles away. It was magnitude -1 on October 7th. As of 2019, it is 13 billion, 485 million miles from the sun traveling at 5,400 mph.

Orbit formula—

$1 = x^2/145^2 + y^2/4.61^2$

perihelion = .578 AU
(9/30/1858)

aphelion = 289 AU
a = 145 AU
b = 4.61 AU

eccentricity = .996

$1/a = 4.610060106 \times 10^{-14}$

period = 1,739 years

Perihelion statistics—

Distance from sun at its closest= 53 million, 714 thousand, 691 miles

Velocity at closest distance from sun= 34 miles, 2,049 feet, 1.49 inches/second

Time of perihelion= 9/30/1858

Aphelion statistics—

Distance from sun at its farthest= 26 billion, 857 million, 345 thousand, 690 miles

Velocity at farthest distance from sun= 447 feet, 3.6 inches/second

Time of aphelion= 2728 AD

Return time= 3597 AD

Comet #54—

Name— Great Comet Tubbutt

Picture comet—

Picture discoverer—

JOHN TEBBUTT, F.R.A.S., 1915.
(Peninsula Observatory, Windsor.)
(Page 316.)

John Tubbutt

Type of comet—elliptical orbit

Comet Tubbutt is a log period comet that was visible for 3 months. It was one of eight great comets in the 19th century. It was discovered by John Tubbutt on 4/3/1861, and on 5/13/1861, it was magnitude 4. Perihelion was reached on 6/12/1861, and it was visible in the Southern Hemisphere starting June 29th. At a distance of 12 million, 331 thousand, 800 miles from earth, it was magnitude 0 to -2, the head very bright and blazing as large as Venus, and the tail was over 90 degrees long, with the earth within the comet's tail. It had a large nucleus and the tail was fan like that reached across 1/4th of the heavens. This comet was comet c/1500 H1 which perihelion on 4/30/1500. It is now more than 9 billion, 300 million miles from the sun.

Orbit formula—
$1 = x^2/180^2 + y^2/16.24495^2$

perihelion= .73455 AU (6/12/1861)

aphelion= 359 AU
a= 180 AU
b= 16.24495 AU

eccentricity= .995919166

1/a= 3.71365953 x 10^-14

period= 2,414 years

Perihelion statistics—

Distance from sun at its closest= 68 million, 263 thousand, 195 miles

Velocity at closest distance from sun= 30 miles, 2,658 feet, 11.75 inches/second

Time of perihelion= 6/12/1861

Aphelion statistics—

Distance from sun at its farthest= 33 billion, 362 million, 585 thousand, 130 miles

Velocity at farthest distance from sun= 384 feet, 4.35 inches/second

Time of aphelion= 3068 AD

Return time= 4275 AD

Comet #55—
Name— Great Comet Coggia
Picture comet—

Picture discoverer— n/a
Type of comet—hyperbolic

orbit

The Great Comet Coggia is a periodic comet which was easily seen by the naked eye in early June of 1874 at magnitude 0 to 1 on July 13th. It was discovered by Jerome Eugene Coggia on 4/17/1874. It was at its brightest on 7/13/1874 at magnitude 0. ``On July 16th, it had a 45 degree length tail which was straight and narrow. Further on, on 23rd July, the tail grew to 70 degrees length as it passed earth at 26 million, 970 thousand miles. The nucleus was as bright as a 1st magnitude star's and the head was 1/2 the size of the moon. The tail was shaky and extending, lighting up and extinguishing like auroras, like a gauze wavering in a strong breeze, pulsating unevenly, some rapid, some longer. An hour later, the tail kindled and extinguished many times, sometimes so no tail could be seen, and sometimes so bright in spite of the moon's light, and all its contours could easily be distinguished to its very extremity.

Orbit formula—

$1 = x^2/572.72^2 + y^2/28.048^2$

perihelion= .676 AU (July 1874)

aphelion= 1,144.7 AU

a= 572.72 AU

b= 28.048 AU

eccentricity= .9988

1/a= $1.167205719 \times 10^{-14}$

period= 13,708 years

Perihelion statistics—

Distance from sun at its closest= 62 million, 722 thousand, 23 miles

Velocity at closest distance from sun= 31 miles, 4,330 feet, 11.477 inches/second

Time of perihelion= 7/1874

Aphelion statistics—

Distance from sun at its farthest= 106 billion, 379

million, 251 thousand, 300 miles

Velocity at farthest distance from sun= 100 feet, 11.56 inches/second

Time of aphelion= 8728 AD
Return time= 15,582 AD

Comet #56—

Name— The Great Southern Comet of 1882

Picture comet—

Picture discoverer— n/a

Type of comet—elliptical orbit

The Great Southern Comet of 1882, discovered on 3/18/1882, became very bright on 9/4/1882 and was a sun grazer, which passed within 1 solar radii of the sun's photosphere at perihelion, so it is a member of the Kreutz family of comets. When it was near the sun at daytime at perihelion on September 7th, it brightened rapidly. It very much looked like a 1st magnitude star at daytime. At perihelion, it was 300,000 miles from the surface of the sun and became magnitude -17 at daytime. (51 times brighter than the full moon.) At post perihelion, it broke into 25 fragments. The nucleus reached maximum size in December 1882, and the tail was visible until February 1883. The last sighting was in June of 1883. This comet is the result of a comet breakup, comet x/1106c1, which also resulted in comet Ikeya Seki and comets du Toit (c/1945x1), 1843d1, 1880c1, 1882r1, 1887b1, 1963r1, and 1965s1, and 1970k1, all of which are sun grazers, the descendants of one comet. The progenitors are either the comet of February 423 or February 467.

Orbit formula—

$1 = x^2/83.7825^2 + y^2/.709^2$

perihelion= .0032 AU (9/7/1882)

aphelion= 167.5617255 AU

a= 83.7825 AU

b= .709 AU

eccentricity= .999964193

1/a= 7.978504016 x 10^-14

period= 772+/-3 years

Perihelion statistics—

Distance from sun at its closest= 297 thousand, 382 miles

Velocity at closest distance from sun= 462 miles, 3,283 feet, .843 inches/second

Time of perihelion= 9/7/1882

Aphelion statistics—

Distance from sun at its farthest= 15 billion, 571 million, 844 thousand, 940 miles

Velocity at farthest distance from sun= 46 feet, 7.775 inches/second

Time of aphelion= 2268 AD

Return time= 2654 AD

Comet #57—
Name— Great Daylight Comet of 1910

Picture discoverer— n/a

Type of comet—elliptical orbit

The Great Daylight Comet of 1910 outshined Venus and was probably the brightest comet of the 20th century. It was discovered on 1/12/1910 at magnitude -1. Perihelion was on January 17th at magnitude -5 and was visible at daytime with a 50 degree long tail by February. It was magnitude -4 on 1/30/1910, and passed closest to earth at 80 million, 600 thousand miles.

Orbit formula—
$1 = x^2/1,487^2 + y^2/4.70234^2$

perihelion= .12875 AU (1/17/1910)

aphelion= 2,974 AU
a= 1,487 AU
b= 4.70234 AU

eccentricity= .999995

1/a= 4.495351146 x 10^-15

period= 57,300 years

Perihelion statistics—

Distance from sun at its closest= 11 million, 985 thousand, 904 miles

Velocity at closest distance from sun= 72 miles, 4,588 feet, 1.67 inches/second

Time of perihelion = 1/17/1910

Aphelion statistics—

Distance from sun at its farthest = 138 billion, 189 million, 872 thousand, 100 miles

Velocity at farthest distance from sun = 46 feet, 5.637 inches/second

Time of aphelion = 30,560 AD

Return time = 59,210 AD

Comet #58—

Name— Great Comet Flaugergues

Picture comet—

Picture discoverer— n/a

Type of comet—elliptical orbit

Great Comet Flaugergues was visible for a record 260 days till Hale Bopp in 1997 took the record. It was discovered on

3/25/1811 by Honore Flaugergues. The comas of the comet was 50% the size of the sun. In August, the comet was in Leo Minor and had 2 tails which resembled a parabola. In October of 1811, it was brightest at 111 million, 600 thousand miles of earth at magnitude 0, with an easily visible coma. On October 6th, the tail was 25 degrees long. By January 1812, it had faded from view. William Blake had witnessed the comet and incorporated it into his panel-'The ghost and the flea'. Tecumseh claimed it a favorable omen concerning the pan native alliance.

Orbit formula—

$$1 = x^2/212.4^2 + y^2/20.94724^2$$

perihelion = 1.04 AU (9/12/1811, 48th century)

aphelion = 423 AU

a = 212.4 AU

b = 20.94724 AU

eccentricity= .995125

1/a= 3.14716724326 × 10^-14

period= 3,096 years

Perihelion statistics—

Distance from sun at its closest= 96 million, 649 thousand, 272 miles

Velocity at closest distance from sun= 25 miles, 3.328 feet, 9.5 inches/second

Time of perihelion= 9/12/1811

Aphelion statistics—

Distance from sun at its farthest= 39 billion, 310 million, 232 thousand, 620 miles

Velocity at farthest distance from sun= 437 feet, 4 inches/second

Time of aphelion= 3359 AD

Return time= 4907 AD

Comet #59—

Name— IRAS Araki Alock

Picture comet—

Picture discoverer—

George Alock

Type of comet—elliptical orbit

Comet Araki Alock is a long period comet which made the closest approach to earth of any comet, 2 million, 9 thousand, 970 miles. It was discovered by George Alock on 4/25/1983 with an absolute magnitude of

12.599. it appeared circular cloud the size of the full moon. It was magnitude 3-4 with no tail, and it traveled 30 degrees per day. Its albedo is .02 and size 5 miles, 3,717 feet in diameter. The comets trail is the cause of the Lyrid meteor shower between Vega and Cygnus at 1-2 meteors per hour.

Orbit formula—
$1 = x^2/98.022^2 + y^2/13.90555^2$

perihelion= .991341 AU

aphelion= 195.0524769 AU

a= 98.022 AU

b= 13.90555 AU

eccentricity= .9898865

1/a= 6.819482735 x 10^-14

period= 970 years, 179 days

Perihelion statistics—

Distance from sun at its closest= 92 million, 127 thousand, 293 miles

Velocity at closest distance from sun= 26 miles, 1,148 feet, 10.82 inches/second

Time of perihelion= 1983

Aphelion statistics—

Distance from sun at its farthest= 18 billion, 126 million, 615 thousand, 230 miles

Velocity at farthest distance from sun= 703 feet, 6.67 inches/second

Time of aphelion= 2468 AD

Return time= 2953 AD

Comet #60—

Name— Comet Arend Rigaux

Picture comet— n/a

Picture discoverer— n/a

Type of comet—elliptical orbit

Comet Arend Rigaux is a periodic comet with a size of 5 miles, 1,360 feet diameter, and its albedo from mars is .0867. It was discovered on 2/5/1951 by Sylvan Arena and Rernend Rigaux.

Orbit formula—
$1 = x^2/3.562^2 + y^2/2.8488^2$

perihelion= 1.423 AU (7/15/2018, 4/10/2025)

aphelion= 5.701 AU

a= 3.562 AU

b= 2.8488 AU

eccentricity= .6003

1/a= 1.876638729 x 10^-12

period= 6 years, 263 days

Perihelion statistics —

Distance from sun at its closest= 132 million, 242 thousand, 225 miles

Velocity at closest distance from sun= 19 miles, 3,301 feet, 7.11 inches/second

Time of perihelion= 7/15/2018

Aphelion statistics—

Distance from sun at its farthest= 529 million, 805 thousand, 286.5 miles

Velocity at farthest distance from sun= 4 miles, 4,744 feet, 6.025 inches/second

Time of aphelion= 2023

Return time= 2020

Comet #61—

Name— Comet Pons Gambert

Picture comet—

Picture discoverer— n/a

Type of comet—elliptical orbit

Comet Pons Gambert is a long period comet discovered 6/21/1877 by Jean Louis Pons and Felix Adolphe Gambert. It was considered a lost comet, but was rediscovered 11/7/2012.

Orbit formula—

$1 = x^2/32.542^2 + y^2/7.21813^2$

perihelion= .81043 AU (12/19/2012, 8/17/2191)

aphelion= 64.274 AU

a= 32.542 AU

b= 7.21813 AU

eccentricity= .97509

$1/a = 2.054141464 \times 10^{-13}$

period= 185 years, 183 days

Perihelion statistics—

Distance from sun at its closest= 75 million, 314 thousand, 874 miles

Velocity at closest distance from sun= 28 miles, 4,691 feet, 9.98 inches/second

Time of perihelion= 12/19/2012

Aphelion statistics—

Distance from sun at its farthest= 5 billion, 973 million, 110 thousand, 855 miles

Velocity at farthest distance from sun= 2.021 feet, 6.02 inches/second

Time of aphelion= 11/7/2105

Return time= 8/17/2191

Comet #62—
Name— Comet Ikeya 1963 1
Picture comet— n/a

Picture discoverer—

Ikeya

Type of comet—elliptical orbit

Orbit formula—
$1 = x^2/95.425634^2 + y^2/10.9643886^2$

perihelion= .632 AU (1963)

aphelion= 190.2180679 AU

a= 95.42564 AU
b= 10.9643886 AU

eccentricity= .993377

1/a= 7.005022524 x 10^-14

period= 932 years

Perihelion statistics—

Distance from sun at its closest= 58 million, 733 thousand, 19 miles

Velocity at closest distance from sun= 32 miles, 4,564 feet, 3.77 inches/second

Time of perihelion= 1963

Aphelion statistics—

Distance from sun at its farthest= 17 billion, 677 million, 343 thousand, 970 miles

Velocity at farthest distance from sun= 577 feet, 24.26 millimeters/second

Time of aphelion= 2429 AD

Return time= 2895 AD

Comet #63—
Name— Comet Neujmin
Picture comet— n/a
Picture discoverer—

Gregory Neujmin

Type of comet—elliptical orbit

Comet Neujmin is a periodic comet discovered by Gregory Neujmin on 2/24/1916. It was last see 2/10/1927 and is considered a lost comet. The comet is still expected to come to a perihelion of 120 million, 900 thousand miles from the sun.

Orbit formula—
$1 = x^2/3.089^2 + y^2/2.164^2$

perihelion= .1.338 AU (3/12/2014, 8/18/2019)

aphelion= 4.84 AU

a= 3.089 AU

b= 2.164 AU

eccentricity= .567

1/a= 2.16400123 x 10^-12

period= 5 years, 157 days

Perihelion statistics—

Distance from sun at its closest= 123 million, 343 thousand miles

Velocity at closest distance from sun= 20 miles, 132 feet, 9.52 inches/second

Time of perihelion= 8/18/2019

Aphelion statistics—

Distance from sun at its farthest= 449 million, 790 thousand, 841 miles

Velocity at farthest distance from sun= 5 miles, 2,829 feet, 4.14 inches/second

Time of aphelion= 5/7/2022

Return time= 2025

Comet #64—

Name— Great Comet of 1680

Picture comet—

Picture discoverer—

Gottfried Kirch

Type of comet—elliptical orbit

The Great Comet of 1680 was the first comet discovered by telescope by Gottfried Kirch on 11/14/1680. It had a very long tail that was visible at daytime, and was 39 million, 60 thousand miles from earth on 11/30/1680. Its perihelion was 580,000 miles from the sun and its peak brightness was on December 29th. It was last seen 3/19/1681. It is, as of February of 2019, 23 billion, 301 million miles from the sun. Isaac Newton used this comet to verify Kepler's laws of planetary motion.

Orbit formula—
$1 = x^2/144^2 + y^2/2.3494^2$

perihelion= .00622 AU (12/18/1680, 12,080 AD)

aphelion= 897.99378 AU

a= 144 AU

b= 2.3494 AU

eccentricity= .999986

1/a= $1.505537647 \times 10^{-14}$

period= 10,400 years

Perihelion statistics—

Distance from sun at its closest= 578 thousand, 37 miles

Velocity at closest distance from sun= 331 miles, 4,353 feet, 7.36 inches/second

Time of perihelion= 12/18/1680

Aphelion statistics—

Distance from sun at its farthest= 82 billion, 523 million, 30 thousand, 880 miles

Velocity at farthest distance from sun= 12 feet, 3.27 inches/second

Time of aphelion= 6880 AD

Return time= 12,080 AD

Comet #65 —

Name — Comet Pons Brooks

Picture comet —

Picture discoverer — n/a

Type of comet — elliptical orbit

Comet Pons Brooks was tentatively identified as the comet of 1313, with sightings in

1240, 1382, 1457, 1529, and 1742, but there was no return as expected in 1670. The comet was discovered in July 1812, then rediscovered in 1883.

Orbit formula—

$1 = x^2/17.1212^2 + y^2/5.0887^2$

perihelion= .77366 AU
(4/21/1924, 5/22/1954)

aphelion= 33.468 AU

a= 17.1212 AU

b= 5.0887 AU

eccentricity= .95481

1/a= 3.904274907 x 10^-13

period= 70 years, 310 days

Perihelion statistics—

Distance from sun at its closest= 71 million, 897 thousand, 765 miles

Velocity at closest distance from sun= 29 miles, 2,190 feet, 9.6 inches/second

Time of perihelion= 5/22/1954

Aphelion statistics—

Distance from sun at its farthest= 1 billion, 592 million, 271 thousand, 600 miles

Velocity at farthest distance from sun= 4 miles, 2,493 feet, 3.84 inches/second

Time of aphelion= 9/23/2059

Return time= 2025

Comet #66—

Name— Great Comet of 1948

Picture comet— n/a

Picture discoverer— n/a

Type of comet—elliptical orbit

The Comet of 1948 had a magnitude of -2 and a tail of 30 degrees length. It disappeared at the end of 1948 in December.

Orbit formula—

$1 = x^2/1{,}931^2 + y^2/21.152723^2$

perihelion= .135 AU (10/27/1948)

aphelion= 3,861 AU

a= 1,931 AU

b= 21.152723 AU

eccentricity= .99994

1/a= 3.461723021 x 10^-15

period= 84,800 years

Perihelion statistics—

Distance from sun at its closest= 12 million, 545 thousand, 819 miles

Velocity at closest distance from sun= 71 miles, 1,186 feet, 6.25 inches/second

Time of perihelion= 10/27/1948

Aphelion statistics—

Distance from sun at its farthest= 358 billion, 810 million, 421 thousand, 200 miles

Velocity at farthest distance from sun= 35 feet, 9.4 inches/second

Time of aphelion= 44,348 AD

Return time= 86,748 AD

Comet #67—
Name— Comet Blanpain

Picture comet— n/a
Picture discoverer— n/a
Type of comet—elliptical orbit

Comet Blanpain is a short period comet discovered by

Jean-Jacque Blanpain on 11/28/1819. It has a very small, disorganized nucleus. It comes into perihelion on 12/20/2019. It is a probable source of the Phoeneid meteor shower.

Orbit formula—
$1 = x^2/2.988^2 + y^2/2.127985^2$

perihelion= .891 AU (8/28/2014, 12/20/2019)

aphelion= 5.294 AU

a= 2.988 AU

b= 2.127985 AU

eccentricity= .702

1/a= 2.237144295 x 10^-12

period= 1,891 days

Perihelion statistics—

Distance from sun at its closest= 82 million, 802 thousand, 405 miles

Velocity at closest distance from sun= 25 miles, 3,033 feet, 2.1 inches/second

Time of perihelion= 12/20/2019

Aphelion statistics—

Distance from sun at its farthest= 2473 million, 395 thousand, 567 miles

Velocity at farthest distance from sun= 4 miles, 2,400 feet, 22.4 millimeters/ second

Time of aphelion= 4/1/2017

Return time= 2025

Comet #68—
Name— Comet Perrine Mrkos

Picture comet— n/a
Picture discoverer— n/a
Type of comet—elliptical orbit

Comet Perrine Mrkos is a periodic comet discovered by Charles Perrine in 12/9/1896. During a perihelion in 1968, it passed 29 million, 230 thousand miles of the earth. The next perihelion is for 1/1/2025, but this comet is considered lost since it has not reappeared since January 1969.

Orbit formula—

$1 = x^2/3.5705^2 + y^2/2.74496861^2$

perihelion= 1.2872 AU (2/26/2017, 1/1/2025)

aphelion= 5.8537 AU

a= 3.5705 AU

b= 2.74496861 AU

eccentricity= .6395

1/a= $1.872171167 \times 10^{-12}$

period= 6 years, 274 days

Perihelion statistics—

Distance from sun at its closest= 119 million, 622 thousand, 60 miles

Velocity at closest distance from sun= 20 miles, 4,669 feet, 6.46 inches/second

Time of perihelion= 2/26/2017

Aphelion statistics—

Distance from sun at its farthest= 543 million, 996 thousand miles

Velocity at farthest distance from sun= 4 miles, 3,128 feet, 5.6 inches/second

Time of aphelion= 7/15/2020

Return time= 1/1/2025

Comet #69—
Name— Comet West Kohoutek

Picture comet— n/a
Picture discoverer—

Kohoutek

Type of comet — elliptical orbit

Comet West Kohoutek os a periodic, Jupiter family class comet discovered by Lubos Kohoutek in late February 1975. On 3/22/1972, it came within 1 million, 116 thousand miles of Jupiter, which reduced its orbital period by 30 years to its current one of 6 years, 175 days sand perihelion from 4.78 AU to 1.6 AU. It made returns in 1987, 1993, 2000, 2006, and 2013.

Orbit formula—
$1 = x^2/3.4707^2 + y^2/2.924056635^2$

perihelion= 1.6012 AU (5/7/2013, 10/26/2019)

aphelion= 5.345 AU

a= 3.4707 AU

b= 2.924056635 AU

eccentricity= .5387

1/a= 1.926005461 x 10^-12

period= 2,365 days

Perihelion statistics—

Distance from sun at its closest= 148 million, 802 thousand, 706 miles

Velocity at closest distance from sun= 18 miles, 739 feet, 3.4 inches/second

Time of perihelion= 10/26/2019

Aphelion statistics—

Distance from sun at its farthest= 496 million, 721 thousand, 497 miles

Velocity at farthest distance from sun= 10 miles, 4,620 feet, 4.03 inches/second

Time of aphelion= 1/1/2023

Return time= 2026

Comet #70—

Name— Comet Honda Mrkos Pajdusakova

Picture comet—

Picture discoverer— n/a

Type of comet—elliptical orbit

Comet Honda Mrkos Pajdusakova is a short period comet discovered by Minoru Honda on 12/3/1948. The color of the comet was green. It is 4,256 feet in diameter.

Orbit formula—
$1 = x^2/3.0205^2 + y^2/1.708746^2$

perihelion= .5296 AU (1/16/1996)

aphelion= 5.511 AU

a= 3.0205 AU

b= 1.708746 AU

eccentricity= .8246

1/a= 2.21367305 x 10^-12

period= 5 years, 91 days

Perihelion statistics—

Distance from sun at its closest= 49 million, 216 thousand, 783 miles

Velocity at closest distance from sun= 34 miles, 1,836 feet, 7.6 inches/second

Time of perihelion= 1/16/1996

Aphelion statistics—

Distance from sun at its farthest= 512 million, 148 thousand, 208 miles

Velocity at farthest distance from sun= 3 miles, 1,594 feet, 3.77 inches/second

Time of aphelion= 10/20/1998

Return time= 2020

Comet #71—
Name— Great Comet of 1769

Picture comet—

Picture discoverer—

Charles messier

Type of comet—elliptical orbit

The Great Comet of 1769 is a long period comet which was visible to the naked eye at magnitude 0. Charles messier discovered it on 8/6/1769 in the constellation of Aries. On September 10th when it was nearest to the earth, the tail was 60 degrees in length. It grew to 90 degrees the next day. The tail was very bright for the first 40 degrees of it, then the rest of it was very dim. The comet was last seen on December 1, 1769.

Orbit formula—
$1 = x^2/163.5^2 + y^2/6.3311405^2$

perihelion= .1228 AU (8/6/1769)

aphelion= 326.8 AU

a= 163.5 AU

b= 6.3311405 AU

eccentricity= .99925

1/a= $4.08843251 \times 10^{-14}$

period= 2,090 years

Perihelion statistics—

Distance from sun at its closest= 11 million, 412 thousand, 49 miles

Velocity at closest distance from sun= 74 miles, 3,517 feet, 11.12 inches/second

Time of perihelion= 8/6/1769

Aphelion statistics—

Distance from sun at its farthest= 30 billion, 370 million, 174 thousand, 990 miles

Velocity at farthest distance from sun= 189 feet, .4 inches/second

Time of aphelion= 2814 AD

Return time = 3859 AD

Comet #72 —
Name — Comet Herschel Rigollet

Picture comet —

Picture discoverer—

Caroline Herschel

Type of comet—elliptical orbit

Comet Herschel Rigollet is a periodic, Halley type of comet discovered by Caroline Herschel on 12/21/1788. The comet returned on 7/28/1939.

Orbit formula—

$1 = x^2/28.843^2 + y^2/6.534324196^2$

perihelion= .74 AU (12/21/1788, 7/28/1939)

aphelion= 56.9 AU

a= 28.843 AU

b= 6.534324196 AU

eccentricity= .974

1/a= 2.317576935 x 10^-13

period= 155 years

Perihelion statistics—

Distance from sun at its closest= 68 million, 769 thousand, 674 miles

Velocity at closest distance from sun= 30 miles, 1,195 feet, 1.79 inches/second

Time of perihelion= 8/9/1939

Aphelion statistics—

Distance from sun at its farthest= 5 billion, 287 million, 830 thousand, 346 miles

Velocity at farthest distance from sun= 213 feet, 3.035 inches/second

Time of aphelion= 2138 AD

Return time= 2094 AD

Comet #73—

Name— Comet Olbers

Picture comet— n/a Picture discoverer— n/a

Comet Olbers is a Haley's type periodic comet discovered by Heinrich Wilhelm Matthias

Olbers on 3/6/1815. Its orbit was calculated by the mathematician Friedrich Gauss on March 31st. The comet returned in 1956, and the next perihelion will occur on 6/30/2024. The nearest approach by the comet to the earth will be 70 million, 308 thousand miles distance on 1/10/2094. The comet has been responsible for a meteor shower on Mars coming from the area of Beta Canis Major.

Type of comet — elliptical orbit

Orbit formula —
$1 = x^2/16.9067774^2 + y^2/6.201520555^2$

perihelion= 1.18 AU (6/30/2024)

aphelion= 32.6351 AU
a= 16.9067774 AU
b= 6.201520555 AU

eccentricity= .93028715

1/a= 3.953791427 x 10^-13

period= 69 years, 183 days

Perihelion statistics —

Distance from sun at its closest= 109 million, 659 thousand, 751 miles

Velocity at closest distance from sun= 23 miles, 3,524 feet/second

Time of perihelion= 6/19/1956

Aphelion statistics—

Distance from sun at its farthest= 3 billion, 32 million, 844 thousand, 853 miles

Velocity at farthest distance from sun = 4,515 feet, 6.234 inches/second

Time of aphelion = 3/19/1991

Return time = 6/30/2024

Comet #74 —

Name — Comet 1999 S4 LINEAR

Picture comet —

Picture discoverer— n/a

Type of comet—hyperbolic orbit

Comet 1999 S4 LINEAR was discovered on 9/27/1999. It came close to earth on 9/22/2000 at a distance of 34 million, 620 thousand miles and reached perihelion on

9/26/2000. It is 1,644 feet in diameter.

Orbit formula— $1 = x^2/(-700^2) - y^2/9.89974^2$

perihelion= .76509 AU (9/26/2000)

aphelion= n/a

a= -700 AU

b= 9.89974 AU

eccentricity= 1.0001

period= n/a (The comet will not return.)

Perihelion statistics—

Distance from sun at its closest= 71 million, 153 thousand, 340 miles

Velocity at closest distance from sun= 98 miles, 4,833 feet, 11.645 inches/second

Time of perihelion= 9/27/2000

Velocity at infinity= 3,692 feet, 11.31 inches/second

Comet #75 —

Name — Comet Shoemaker levy

Picture comet —

Picture discoverer—

Levy

Shoemakers

Type of comet — elliptical orbit

Comet Shoemaker Levy 9, discovered 3/24/1993, was a comet that collided with Jupiter at a speed of approximately 37 miles per second on July 16-22, 1994 and was the first such extraterrestrial collision ever observed. The comet was 1 mile, 528 feet in diameter.

Orbit formula —
$1 = x^2/2.82843^2 + y^2/1.41^2$

perihelion= n/a
aphelion= n/a
a= 2.82843 AU
b= 1.41 AU
eccentricity= .9986
1/a= $1.926005461 \times 10^{-12}$

period= 2 years

Comet #76—

Name— Slaughter Burnham

Picture comet—

Picture discoverer—

Robert Burnham

Type of comet—elliptical orbit

Comet Slaughter Burnham is a periodic comet discovered by Charles slaughter and Robert Burnham on 12/10/1958.

Orbit formula—
$$1 = x^2/5.036^2 + y^2/4.777565^2$$

perihelion= 2.535 AU
(8/4/1958, 7/18/2018)

aphelion= 7.679 AU
a= 5.036 AU
b= 4.777565 AU

eccentricity= .5036

1/a= 1.327360436 x 10^-12

period= 2 years, 197 days

Perihelion statistics—

Distance from sun at its closest= 235 million, 755 thousand miles

Velocity at closest distance from sun= 14 miles, 1,153 feet, 10.77 inches/second

Time of perihelion= 12/19/2027

Aphelion statistics—

Distance from sun at its farthest= 714 million, 147 thousand miles

Velocity at farthest distance from sun= 4 miles, 3,185 feet, 1.39 inches/second

Time of aphelion= 4/25/2024

Return time= 2021

Comet #77—
Name—Eclipsing Comet

Picture comet—

Picture discoverer— n/a

Type of comet—elliptical

The eclipsing comet of 1948 was discovered during a solar eclipse on 11/1/1948 and was magnitude -2 then. It reached perihelion on 10/27/1948. At one point, its tail was around 30 degrees long. It fell beyong naked eye visibility at the end of december.

Orbit formula—
$1 = x^2/1962^2 + y^2/19.61976^2$

perihelion= .135 AU (10/27/1948)

aphelion= 3861 AU
a= 1962 AU
b= 19.61976 AU

eccentricity= .99995
1/a= $3.407027052 \times 10^{-15}$

period= 34,800 years

Perihelion statistics—

Distance from sun at its closest= 12 million, 555 thousand miles

Velocity at closest distance from sun= 71 miles, 1,184 feet, 4.45 inches/second

Time of perihelion= 10/27/1948

Aphelion statistics—

Distance from sun at its farthest= 359 billion, 73 million miles

Velocity at farthest distance from sun= 3,738 feet, 7.16 inches/second

Time of aphelion= 44,348 AD

Return time= 86,748 AD

Pictures of surfaces of a comet

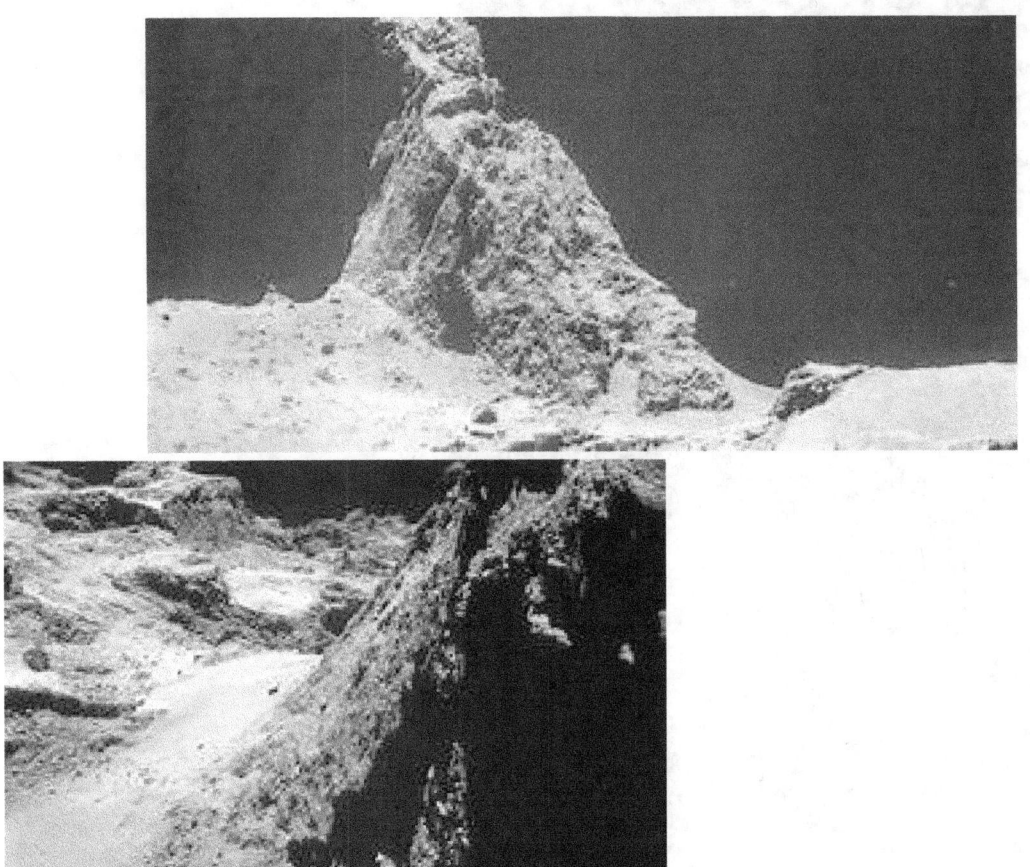

surface pictures of comet 67p/ Chuyumov-Gerasimenko

Orbital calculations—
These are the formulas to compute a comet's orbit—

Will need following informations—

1 mile=1.60934743 kilometers

1 kilometers= .621369867 miles

Speed of light=

186,282 miles/second

299,792,458 meter/second

1,582,502,328 feet/second

670,615,200 miles/hour

1 astronomical unit=

92,955,807.3 miles

149,598,189.6 kilometers

$GM = 1.3274586 \times 10^{20}$

(Note — All numbers are to be in meter, kilograms, and seconds.)

Formulas for orbital calculations

Elliptical orbits—

$1 = x^2/a^2 + y^2/b^2$

Eccentricity between 0 and 1

$c = sqrt(a^2 b^2) = a*eccentricity$

(c is the distance between then very center of the ellipse, which is a flattened circle, to the focus inside one end of the ellipse.)

eccentricity=(a-perihelion)/ a=sqrt(1-(b^2/a^2))

(Eccentricity can be: 0 for a circle, greater than 0 and less than 1 for an ellipse, 1 for a parabola, and greater than 1 for a hyperbola. Eccentricity indicates how flattened the circle is. The greater the number, the more flattened the circle becomes, and at 1 and above eccentricity, the orbit is no longer a closed loop, but opens up like the inside of a wine glass shaped.)

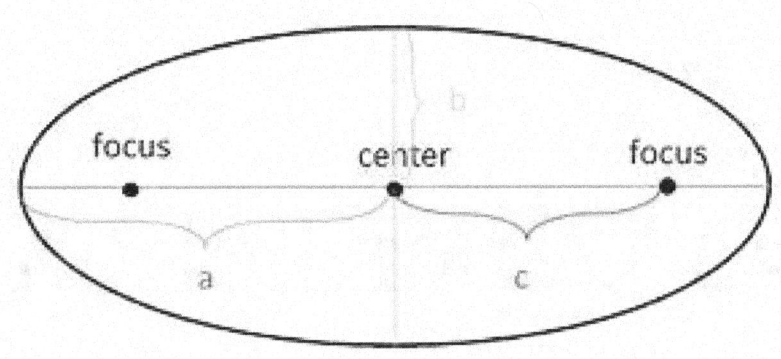

perihelion=a-c=a*eccentricity=a(1-eccentricity)

(Perihelion is the distance from the sun to the end of the flattened ellipse when that distance is the least amount.)

aphelion=a*2-perihelion
(Aphelion is the distance from the sun to the end of the flattened ellipse when the distance is the greatest.)

a=sqrt(b^2+a^2)=perihelion/(1- eccentricity)

(a is the distance from the center of the ellipse to the end of the fattened ellipse.)

$b = \sqrt{a^2 - (\text{eccentricity} \cdot a)^2}$

(b is the distance from the center the ellipse to the nearest part of the ellipse.)

Semi latus rectum = $a(1-\text{eccentricity}^2)$

Period = $a^{(3/2)}$

a = $\text{period}^{(2/3)}$

Area inside orbit = $2\pi \cdot a \cdot b$

perimeter = $2\pi \cdot \text{sqrt}((a^2+b^2)/2)$

velocity = sqrt((1.3274586 x 1-^20)*(2/ current

distance of comet from sun-1/a))

Current distance from sun=

2/((velocity^2)/1.3274586 x 10^20)+1/a)

To find current distance object in orbit from the sun—

1. Find number of years including decimal points since perihelion.

2. Make good guess distance object from sun.

3. Use formula — velocity=sqrt((1.3274586 x 10^20 x (2/ distance guess converted into meters-1/a))

4. Distance guess/ velocity=time since perihelion

5. Times in numbers 1 and 4 match?

6. If times don't match close enough, do steps 2 to 5 again till satisfied times match close enough.

7. Times x velocity=current distance from sun.

Hyperbolic orbit formulas —

eccentricity > 1

$1 = x^2/a^2 - y^2/b^2$

perihelion = $a(1-e)$

eccentricity = c/a = $\sqrt{1 - b^2/a^2}$

$c = \sqrt{a^2 + b^2}$ = eccentricity * a

$a = \text{perihelion}/(1-e) = \sqrt{b^2/(e^2-1)}$

$b = a \cdot \sqrt{e^2-1}$

Semi latus rectum $= a(e^2-1)$

Periapsis distance $= a(1-e)$

Velocity perihelion $= \sqrt{(1.3274586 \times 10^{20})(1+e)/(e-1)}$

Velocity at infinity = sqrt(1.3274586/(-a))

Finding the elliptical orbit equation from 2 sightings —

Coordinates 2 sightings — (5,2), (10,3)

a^2b^2 = 5^2b^2 + 2^2a^2

$a^2 b^2 = 10^2 b^2 + 3^2 a^2$

Subtract —

$75 b^2 = 5 a^2$

$15 b^2 = a^2$

$b = .2583a$, $b^2 = .0666 a^2$

$a = 3.873b$, $a^2 = 15b$

$15 b^2 \times .0666 a^2 = 25 \times .0666 a^2 + 4 \times 15 b^2 =$

$a^2b^2 = 1.665a^2 + 60b^2 =$

$1.665a^2 + 60b^2 =$

$1.665a^2 + 60 \times .0666a^2 = 5.661a^2$

$15b^2 = 84.9157$

$b = 2.3793$

$5.66a^2 = 25 \times 5.66 + 4a^2$

$1.66a^2 = 141.527$

$a = 9.2335$

Equation of orbit —

$1 = x^2/9.2335^2 + y^2/2.3793^2$

Calculating an orbit for a hyperbolic comet from 2 sightings —

Coordinates —

$(10, 2.45), (20, 4.975)$

$1 = 10^2/a^2 - 2.45^2/b^2$

$1 = 20^2/a^2 - 4.975^2/b^2$

Subtract--

$a^2 b^2 = 100 b^2 - 6 a^2$

$a^2 b^2 = 400 b^2 - 24.75 a^2$

$0 = -300 b^2 + 18.74 a^2$

$b^2 = .0625 a^2$

$b = .25 a$

$1 = 10^2/a^2 - 6/.0625a^2$

$1 = (6.25a^2 - 6a^2)/.0625a^4$

$.0625a^4 - 6.25a^2 + 6a^2 = 0$

$.0624a^2 - .25 = 0$

$a^2 = .25/.0625$

$a = 2$

$1 = 100/2^2 - 2.45^2/b^2$

$24 = 6/b^2$

$b = .5$

Orbital formula —

$1 = x^2/2^2 - y^2/.5^2$

Note — will know from calculating an elliptical or hyperbolic orbit from the 2 coordinates which orbital type it is because during calculations, the wrong orbital kind will have a negative square root in it, so it will be the other kind of orbit in that case. (A negative square root is not possible in the orbital calculations cases.)

Glossary

Albedo—The fraction of light reflected from an astronomical body.

Apparent magnitude—How bright an astronomical body appears to the eye. Higher negative numbers are brighter and higher positive numbers are dimmer. Magnitude 5 is about the limit of being able to observe a sky object on a dark, moonless night.

Absolute magnitude—The apparent magnitude of an object if viewed from 32.6 light years away.

Period—How long it takes to travel around the orbital distance one time.

Focus—Where the sun is located inside the ellipse, at either end of the flattened ellipse.

Elliptical orbit— A flattened circular shaped orbit.

Hyperbolic orbit— An open ended path where the orbit is not a closed circuit.

Kepler's 3 laws of planetary motion—

1 The orbit of a planet is an ellipse with the Sun at one of the two foci.

2 A line segment joining a planet and the Sun sweeps out equal areas

during equal intervals of time.

3 The square of the orbital period of a planet is directly proportional to the cube of the semi-major axis of its orbit.

Jupiter family comet— A comet with an orbital period of less than 20 years and is influenced by Jupiter's gravitational pull.

Roche limit—The distance a body from a second body that will cause the break up caused by a second bodies gravitational influence overcoming the first bodies gravity holding the first one together.

GM—The product of the gravitational constant (6.674×10^{-11} m^3 × kg^{-1} × sec^{-2}) and the mass of the sun (1.989×10^{30} kg). Equals 1.3274586×10^{20}.

Kupier belt—Similar to the asteroid belt, it is composed of astronomical bodies, such as comets, left over from the solar system's formation. It is shaped like a donut.

Oort Cloud—Predominantly icy planetesimals surrounding the sun at a distance range of .3 to 3.2 light years.

Kreutz Comets—Comets that come extremely close to the sun at perihelion. (Sun grazing comets)

Periapsis—Distance closest to to focus of an elliptical orbit.

www.ingramcontent.com/pod-product-compliance
Lightning Source LLC
Chambersburg PA
CBHW082335230526
45466CB00023BA/2954